"Fascinating"

—*Daily Mail*

"Excellent. . . a monumental achievement in data gathering. . . a funny and entertaining read. . . essential reading before you place your bets in the big lottery of life!"

—*BBC Focus*

"The statistics are presented with admirable lucidity, using an ingenious method devised by the authors, and may serve to reassure more neurotic readers. . . . Witty and illuminating, *The Norm Chronicles* is essential reading for anyone wanting to know whether they should try skydiving, or accept that third glass of wine."

—*Financial Times*

"Blastland and Spiegelhalter wear their learning lightly. They never lapse into the smug tone that often accompanies books like these. . . . Helping people make sense of the barrage of confusing (and often misrepresented) statistics that riddle every day is a noble goal. Making the process enjoyable is a real achievement."

—*The Economist*

"A witty, insightful, educational and wholly original book—and a wonderful achievement. Read it!"

—Tim Harford, author of *The Undercover Economist*

"Blastland & Spiegelhalter achieve the amusing feat of bringing mortality data back from the dead. Reading this book will, in all probability, add years to your life."

—Kaiser Fung, author of *Numbers Rule Your World* and *Number Sense*

"We have a really tough time understanding absolute, individualized risk—until now. *The Norm Chronicles* provides a long overdue, systematic, and entertaining dissection of life's risks."

—Eric Topol, M.D., author of *The Creative Destruction of Medicine*

"Accessible yet deep, *The Norm Chronicles* explains how statistical regularities and irregularities are central to every aspect of our lives. If Jonathan Coe and Gerd Gigerenzer were to collaborate on a sardonic self-help book, this is what it might look like."

—Andrew Gelman, Professor of Statistics and Political Science, Columbia University

"Numbers matter, especially in the face of risk. This book is a powerful remedy for a deadly affliction—innumeracy."

—Paul Slovic, president of Decision Research, and author of *The Feeling of Risk*

"Blastland and Spiegelhalter's *The Norm Chronicles* is irreverent, poignant, insightful, and just about the best book about risk I've ever read. It's also a paradox—a book about numbers and probabilities that'll keep you hooked to the last page. That shouldn't be possible. Using master storytelling and a large dose of humanity, Blastland and Spiegelhalter transform the statistics of danger and death into a celebration of life. It's a rare feat, but one that's as compelling as it is important. This book is essential reading to anyone who has ever faced the possibility of something going wrong, and thought 'What the …?!' Buy it!"

—Andrew Maynard, director, University of Michigan Risk Science Center

"In the same manner that the bumblebee disproved the calculations of an earlier time's aerodynamics, so Blastland and Spiegelhalter refute a central tenet of today's science of risk communication: that the meaning of numbers defies the narrative currency of everyday reasoning. Engaging, enlightening stories of probability, they demonstrate, are the most reliable means for transmitting empirical knowledge of the dangers we face and how to abate them."

—Dan Kahan, Professor of Law and Psychology, Yale Law School

THE NORM CHRONICLES

Despite starting out telling stories as an English literature graduate working in journalism, MICHAEL BLASTLAND somehow learned to count, devising *More or Less*, the BBC Radio 4 program about numbers, and writing, with Andrew Dilnot, *The Tiger That Isn't*, a guide to numbers in the news. On ice skates, he is life threatening (mostly it's his own life), but he is more afraid of other people's raised umbrellas. His other risk dislikes include confined spaces, heights, and fairground rides.

DAVID SPIEGELHALTER is a statistician who rejoices in the title of Winton Professor for the Public Understanding of Risk at Cambridge University. He is, or was, a proper academic, and has far too many letters after his name, but feels his greatest achievement is not doing badly in the risky BBC TV program *Winter Wipeout*. He lives in a flood zone, but is more anxious about forgetting where he put the house keys. He particularly likes heights, confined spaces, and fairground rides.

Together, Blastland and Spiegelhalter sound like a Dickensian music-hall act or a firm of dodgy lawyers. But they think that combining ideas and perspectives is the best way to make sense of the subjects of this book: danger, risk, and chance—subjects that could be said to exist, as they hope to show, only in a clash of viewpoints.

THE NORM CHRONICLES

STORIES AND NUMBERS ABOUT DANGER AND DEATH

Michael Blastland and
David Spiegelhalter

BASIC BOOKS
A Member of the Perseus Books Group
New York

Books published by Basic Books are available at special discounts for bulk pur-
chases in the United States by corporations, institutions, and other organizations.
For more information, please contact the Special Markets Department at the Perseus Books Group, 2300 Chestnut Street, Suite 200, Philadelphia, PA 19103, or call
(800) 810-4145, ext. 5000, or e-mail special.markets@perseusbooks.com.

Designed by Jack Lenzo

Library of Congress Cataloging-in-Publication Data
Blastland, Michael.
  The Norm chronicles : stories and numbers about danger and death / Michael
Blastland and David Spiegelhalter.
    pages cm
Bibliographical references and index.
  ISBN 978-0-465-08570-5 (paperback) -- ISBN 978-0-465-08569-9 (e-book)
1. Risk--Sociological aspects 2. Accidents--Statistics. 3. Disasters--Statistics. 4.
Violent deaths--Statistics. I. Spiegelhalter, D. J. II. Title.
  HM1101.B53 2014
  363.1--dc23
                                    2014004656

10 9 8 7 6 5 4 3 2 1

For Ron and Shirley—MB

For Kate, Kate, and Rosie,
for putting up with me—DS

CONTENTS

INTRODUCTION

A short story about danger:

One day, by devilish coincidence, three people traveling separately on the subway—Norm, Prudence, and Kelvin—saw three unattended bags.

For Norm it was a dusty-blue canvas tote on the floor, tucked against the seat on his side. At first, he thought nothing of it. Then he looked again, and then up and down the carriage. Almost empty.

"Calm, Norm," he said to himself, and reached down to pull up his socks, glancing sideways for wires. He sat up and forced himself to focus on the probabilities, scratched his nose, concluded several times that the bag had been forgotten, that's all, stood up and went slowly to the far doors to get off at the next stop and enjoy the extra walk.

When Prudence looked up from *Fifty Shades of Grey* to see a pristine backpack on the seat opposite, she felt quickly sick. If it has a travel tag, someone is taking it somewhere. If it doesn't . . .

It didn't. She thought of her children, motherless, crying, and lost the strength to move. Her mind filled with hellish images of herself blown apart and her hair ruined.

Counting out the last seconds of her life, ticking to the blast, she gestured and mouthed at a passenger standing nearby, a warning, a hope: "a . . . bag," she mumbled, and pointed like the ghost she was becoming.

"Oh, yeah," he said, and grabbed it. "Cheers."

And Kelvin? When he clocked a black briefcase as the doors slid open, he opened it—what else? Picked it up, sat down, flipped up the sucker's lid, took out a folded *Daily Telegraph* and slid it into the side-pocket of his leather jacket, shuffled through a wad of paperwork, noticed a foil-wrapped packet, teased it open, lifted it to his nose, snorted—with an eye on the teenage girl further down, doing her lashes—threw in the wrapper, closed the lid, put down the bag, sat back, and closed his eyes.

Three people, three points of view about danger, to which we could add many more. What's yours? Danger, as every experience and a million stories tell us, is in the nerves of the beholder.

But not entirely. There are also numbers.

Here are two. The first is terrible and well known: on July 7, 2005, 52 people died in terrorist bomb attacks on underground trains and a bus in central London. Second, in 2011, about 30,000 bags were left on London's buses and subways. Let that sink in: 30,000.

So, is an unattended bag on the subway dangerous? How do the numbers and stories compare: the particular stories of the 52, and those of Norm, Kelvin, and Prudence?

Leave that question burning for a moment, for another short story, a true and famous one.

One day Anna and her friends went skiing. Anna was a good skier, and in any case there's not much chance that skiing will kill you. Then she lost control. She fell onto her back on a frozen stream near a waterfall. A hole opened in the ice. Freezing water poured into her clothes and pulled her in, head first.

She would have been dead in minutes, but under the ice she found an air pocket. She could breathe, for a while. First her friends, then a rescue team, tried to pull her out. They failed. They tried to dig her out. The ice was too hard.

Anna remained conscious for forty minutes. But finally her breathing slowed, and then stopped. Then so did her pulse. It was another forty minutes before she was pulled clear.

Normal body temperature is 37° Celsius. Hypothermia begins to set in at about 35°C. When Anna arrived at the hospital, her temperature was 13.7°C. No one that cold had ever lived.

But the doctors did not give up. Slowly, patiently, they warmed her blood outside her body, then pumped it back into her veins. More than three hours after she stopped breathing, more than two hours after she arrived at the hospital, Anna's heart began to beat again.

But when she woke, ten days later, it was to find herself paralyzed from the neck down, and one of her first feelings was anger that she had been revived, for this. Then she made an almost full recovery. A few years later she was working as a radiologist at the hospital that saved her life. She still goes skiing.[*]

Anna's story is now celebrated as a marvel of survival and a medical revelation. But for us it makes a different point that has little to do with human endurance or scientific understanding of the effects of extreme cold, and is simply this: Anna rode the mother of all fortune's roller coasters. At every turn in a twisting tale, at every roll of the dice for good or bad, it was as if she threw six 6s.[†] All her luck was extreme.

Anyone can fall, even those who ski well. But how and where Anna fell—into that concurrence of water, hole, and hard ice—was absurdly unlikely. Then, to find an air pocket was a godsend, or seemed so, but for the desperate twist that made it so hard to pull her out. To be visibly dead from such extreme cold and yet live was impossible, until she did it; then to survive but in the end wake up paralyzed, except that the end was still to come with almost full recovery, was one amazement after another. And behind every turn was the most twisted of all: that it was near-death that saved her, when fatally bitter cold turned out—by slowing her metabolism

---

[*] The story of Anna Bagenholm comes from various sources, including M. Gilbert, R. Busund, A. Skagseth, P. Å. Nilsen, and J. P. Solbø, "Resuscitation from Accidental Hypothermia of 13.7 Degrees C with Circulatory Arrest," *The Lancet* 355, no. 9201 (2000): 375–376, and Atul Gawande, *Better: A Surgeon's Notes on Performance* (London: Profile Books, 2008).

[†] The usual phrase here is that Anna beat or defied the odds. Strictly speaking, no one can defy the odds. Odds simply describe how many people are expected to be on each side of a possibility. Even when the odds are a million to one against something happening to someone, when you turn out to be the one, this is not beating the odds: it is *meeting* the odds.

down nearly to a standstill at just the right moment—to preserve a kernel of life precisely when breath stopped. Life is sometimes improbable.

What our brief sample of stories and numbers tells us is that risk is two-faced: on one side are the seemingly hard-nosed calculations of probability, such as the 20 percent extra risk of cancer from eating sausages that hits the headlines, or the infinitesimal percentage of bags, unattended or otherwise, that explode in London, or the chance of surviving if your body freezes, you stop breathing, and your heart quits; on the other side are people and their stories, like Anna, or the 52.

Numbers and probabilities tend to show the final account, the risks to humans en masse, chance in aggregate summarized for whole populations. These numbers reveal hypnotic patterns and rich information. But they are indifferent to fate and its drama. Numbers can't care and don't care; life and death are percentages, unafraid of danger, shrugging at survival, stating only what's risky, what's not, or to what degree, on average. They are silent about how much any of this, right down to a love or fear of sausages or ski slopes, matters.

But we—and you—are not averages. We are also subjective: we do care and might even argue about skiing, terrorism, and sausages. We have our instincts, feelings, hopes, fears, and confusions. Our intuition might not match the stats, and we might say, "So what, I'm out of here." Or maybe we see danger and take the leap anyway, base-jumping from a cliff edge in a wingsuit because we love the buzz (see extreme sports, in Chapter 16); maybe we run screaming from spiders (phobias appear in Chapter 25). We ask, "Will I be safe? Will my children be safe?" But also "Will I be in control?" (Chapter 15), "Will I be happy?" "Do I want this thrill?" "Do I value this choice?" (drugs are covered in Chapter 9), "Should I take this chance?" And "How does it feel, and what's it worth?" (see childbirth, in Chapter 11).

An extreme illustration of the difference is that, as far as the mortality statistics go—the statistics used to calculate the risk of skiing, along with many of life's other hazards—nothing happened to

Anna. She was a tick, not a cross; she lived and didn't die, and that's it, all that the mortality record has to say.

Danger is the shark in shallow waters, the pills in the cupboard, or a grand piano teetering on a window ledge while children skip below. It is the diet too rich in cream, the base-jump, the booze, the pedestrian and the double-decker, driving a car fast, or the threat of weirder weather. It is the spills and the thrills. In other words, danger is everywhere and always. And in all cases we find those same two faces: one impassive, formal, calculating, the other full of human hopes and fears.

The unusual aim of this book is to see both at once. We hope to show people and their stories *and* the numbers, together. We set out to do this mainly to explore how these two perspectives compare, but along the way we found that this aim raised an awkward question: Are the two faces of risk compatible? Can risk claim to be true to the numbers and to you at the same time? We will present both sides as we try to find out, but we will tell you our conclusion now.

It can't. For people, probability doesn't exist.

That's an extraordinary claim from writers sometimes geeky enough to have two hoods on their anoraks. But with a little luck, the proof of it—and exactly what it means—will emerge through the clash of perspectives in this book.

The numbers and probabilities are all here. With them we show the chances of a variety of life's tricks and traps: risks to children; risks of violence, accidents, and crime; dangers from sex, drugs, travel, diet, and lifestyle; risks of natural disaster; and more. We say how we know these risks, why sometimes we can't know them, and how they've changed, and we use the best methods we can find or invent to make them easy to grasp. In particular, we use a cunning little device called the "MicroMort," and a new one called the "MicroLife," two friendly units of deadly risk that we think offer real insight. You will meet them soon enough. In this respect, the book is a new guide to life's odds.

The human factor is here, too. People don't always do what the numbers seem to suggest they should. Some feel safe when they are in danger, or in danger when they are safe, and the numbers may matter less to us than feelings of power or freedom, our values, our likes and dislikes, and our emotions.

One reaction to the difference is to tell people they are stupid, and that if only they listened to the experts they'd live longer and sleep sounder. Another is to say that the experts may be right about the averages but that they clearly never had kids or an undiagnosed chest pain, or itched to take a corner too fast.

Either way, the human factor can't be ignored. To show it, we use a technique that is, well, risky: we combine fact with fiction, numbers with stories. Why write a book that's part numbers and part stories? Because that is how people see risk—through both stories and numbers.

Each has its virtues and shortcomings. Numbers tell us the odds. Stories are how we often convey the feelings and values that numbers cannot, feelings and values that might, in turn, distort our perception of the odds. Stories impose order, but often artificially—beginnings, middles, and ends, all tied neatly together (too neatly?) with cause and effect. Numbers give us probabilities, which often don't claim to know the precise causes and effects of how one thing leads to another but simply show us how it all adds up into a tally of life and death. To understand how these perspectives play out, shouldn't we see them together? To be true to them, shouldn't we try to let each speak on its own terms?*

Steven Pinker wrote in *How the Mind Works*:

---

* What is a story? David Herman, ed., in *The Cambridge Companion to Narrative* (Cambridge, UK: Cambridge University Press, 2007), says that "stories are accounts of what happened to particular people—and of what it was like for them to experience what happened—in particular circumstances and with specific consequences" (p. 3). That'll do for us. It's the "particular people" bit that matters, which includes both fact and fiction. Specialists use the word "narrative" to distinguish the way a story is told from the pure events, but this is not a book about narrative theory, although here and there we do talk about the story form—heroic medical stories in Chapter 23, for example.

Fictional narratives supply us with a mental catalogue of the fatal conundrums we might face someday and the outcomes of strategies we could deploy in them. What are the options if I were to suspect that my uncle killed my father, took his position, and married my mother? If my hapless older brother got no respect in the family, are there circumstances that might lead him to betray me? . . . What's the worst that could happen if I had an affair to spice up my boring life as the wife of a country doctor? . . . The answers are to be found in any bookstore or any video store.[1]

So we created some characters. First, Norm, the one who saw the blue tote bag on the subway and tried to compute the optimal, proportionate reaction: our hero. Something or someone is out to get Norm, though he's done nothing wrong. He's just an average guy (the clue is in the name), looking for a safe path through life. He's so average that even his attempts to stand out are average, the sort of guy who feels moderately about licorice.

But life has its own plans for Norm: maybe a car crash, a fatal dose of bird flu, a mugger's knife or meteorite, a nuclear meltdown or his own spreading waistline. Somewhere, an assassin waits.

Still, he tries. Risk is calculable, he says, and with reason as his guide and a sense of proportion, he can steer a true course. His habits are ordinary: he likes a nice cup of tea but not too many, wears off-the-rack menswear, and invites little risk from hot passion or daring. Even so, someone or something wants him dead. Norm's entire, blameless life is a story of mortal danger, as to some extent is yours, and ours.

Prudence (another clue), the one who panicked on the tube, treads warily, all anxious glances. Every stranger's footstep could be following her. The numbers hardly matter, and one scary story is all that fear needs to set fire to her imagination.*

---

* There's a tendency for women to be more risk-averse than men, but only on average. We are aware that we play with stereotypes all round.

Finally, there are the Kevlin brothers, Kelvin, Kevin, and Kieran, chancers and risk-junkies who fly by the seat of their pants and might, just for the hell of it, tell you exactly where to shove your reasons and probabilities.

Side by side, chapter by chapter, numbers and stories. We had planned to let the clash of perspectives stand on its own without comment. In the end we went a step further to bring out the differences. So within the nonfiction realm we also explore the psychology of risk perception. This is where numbers and stories meet, and often disagree.

All of which leaves us with two competing worldviews. What happens when they collide? A fight, often as not, in which advocates of one approach accuse the others of being irrational, and the others reply that the first lot are unfeeling.

These conflicts are elemental. Within people's attitudes to risk lurk many of life's deepest tensions. Pick your side: art versus science, feeling versus reason, words versus numbers, perception versus objectivity, stories versus stats, instinct versus analysis, the particular versus the abstract, romanticism versus classicism, red trainers versus brown Hush Puppies: in short, the eternal row between fundamentally different versions of truth and experience. It is easy to set up camp in one or the other and never look outside.

Even if you think you don't take these sides, and certainly not so crudely, you might take them over danger. The choice touches something deep in our attitude to life; it helps define what kind of people we are. Sometimes we embody both sides and are torn by contrary impulses as the struggles between different worldviews—sometimes the numbers, sometimes the stories and emotions—ebb and flow while we try to work out how to live. Beside such weighty stuff, the fact that danger might also strangle us with a blind cord, poison us with salmonella, or blow us to bits seems almost by the way.

Here's a quick illustration.

It's a summer's day. You are walking down the street licking a Good Humor cone when a number 42 bus thumps onto the sidewalk, whacks your ice cream from your hand, and rips into 7-Eleven in a blizzard of glass and twisted metal, leaving you shocked—but unharmed. What are the chances?

In no instance are the exact facts of an accident predictable with anything like certainty. True, we know, to take another example, that leaning out from a ladder to paint the irritating bit in the far corner is asking for it. We can see what's coming, we say. But can we? Reliably? Of course not. Sometimes it comes, and sometimes it doesn't. Chance always plays a part.

Similarly, a bus crash might be mechanical failure or driver error; it might depend on the timetable, the road, the traffic, on the weather, on you, on the length of the queue for the ice cream, which in turn depends on everyone in it, or it might depend on every one of those things—in all, an infinitely intricate and improbable spaghetti of causes, events, and people. You, reading this word in this book at whatever time and place you happen to be, took an absurd quantity of cause and effect—right back to the beginning of everything, if you must. Which is another way of saying that no one knows the future. Life is too complicated.

And yet you might not be all that surprised if a crash like that really happened—if not to you, then to someone. We know for sure that countless things—unlikely or not—will happen somewhere to someone, as they must. More than that, we know that they will often happen in strange and predictable patterns. Fatal falls from ladders among the approximately 21 million men in England and Wales in the five years to 2010 were uncannily consistent, numbering 42, 54, 56, 53, and 47. For all the chance particulars that apply to any individual among 21 million individuals, the numbers are amazingly, fiendishly stable—unlike the ladders.[2] Some calculating God, painting fate by numbers up in the clouds, orders another splash of red: "Hey, you in the dungarees, we're short this month."

We know there will be accidents and incidents, and we often know what kind and how many—to the extent that we can predict pretty well how many people will be murdered in London by July 28 next year, and even how many murders there are likely to be on one day (which we did—in Chapter 22, on crime). Up close, life can appear chaotic. Every murder is unique and unpredictable, every fall and crash laced with infinite chance. Seen from above, people often move in patterns with spooky regularity.

This is the great puzzle about danger: that a million stories describe it, feelings inform it, and a million occasions conspire for or against every incident—and yet there are relatively consistent numbers. Every cancer begins with a freak cellular mutation, and yet a fairly steady one-third of people will get one.* It is one of life's odder facts, this order amid disorder, the natural and spontaneous emergence of shape—persistent and predictable—even as individuals do their own thing.

So, from above, the course of human destiny is often clear. To individuals below, it is a maze of stories. It is as if there were two forces at the same time: one at the big scale pulling toward certainty, the other pushing individuals toward uncertainty. There's a word to describe this balance between the patterns of populations and the stumbling of a single soul, a word first used in its modern sense only a few hundred years ago: probability.

Probability—at least one version—begins with counting past events, such as in "20 percent of the men who died in recent years died from heart disease." It then uses this fact to predict a pattern. "About 20 percent will die from heart disease in the future." But then it goes a step further: from that general prediction it gives odds for what will happen to individuals. "The risk or chance that the average person will eventually die from heart disease is therefore also 20 percent, or one in five, for a man, and about 14 percent, or one in seven, for a woman." Thus it moves from past to future, from the mass to the individual.

Probability is magical, a brilliant concept. It yokes together our two worldviews, the two faces of risk: the orderly view of whole

---

* Although, unlike falls from ladders, some numbers do change dramatically if some big causal factor has also changed. For example, the incidence of heart disease has declined, largely because people are smoking less. Deaths from heart disease in men in the United Kingdom fell from 147 per 100,000 people in 2005 to about 108 per 100,000 in 2010, and for women from 69 to 48 per 100,000. These are huge changes.

populations seen in numbers from above and the sometimes lonely view in the maze of stories below. It embraces all of us as individuals in aggregated data. Today people use principles of probability in making decisions about everything from the weather to money, or to calculate their chance of being burgled, or to decide how much risk they are taking when using cellphones, eating sausages, or going to places that are sometimes hit by tsunamis. Probability touches our hopes and fears at every turn. The news is full of it, and no wonder—it seems to offer a hold on the future. Which is why it is a little bit inconvenient, as we shall argue later, that it doesn't actually exist out there in the world.

Norm's life-course shapes the order of the risks here, as we discover how well numbers can guide him, from beginning to end. As for which to include, we chose whatever seemed interesting and personal. There's little, for example, about risks in business—or Enterprise Risk Management (ERM), as it is known—about which more than enough is said already.

To add to our ambitions, we hope that the different perspectives here create a book that everyone from all sides of the argument can read and enjoy on the path to mutual understanding. Although by trying to reach everyone, it has occurred to us that we might reach no one. It's a risk.

Finally, a couple of hazard warnings: first, people are learning more about risks all the time, so it won't be long before there will be new data. This fact is not just awkward, it's relevant to the argument. What kind of trust can you place in numbers that don't stand still?

Second, in places this is almost a mini-encyclopedia of hazards. There are a lot of numbers here, and they are more fun to dip into than to read in one go. Even so, the statistical evidence could go on and on, and we welcome readers' arguments about all those extra stats that we left out. We had to stop somewhere.

So the scene is set. Stories will be pitched beside stats, reason will tangle with feeling and impulse, belief will quarrel with evidence. We looked hard at the data and threw them at an imagined life where objectivity doesn't always get a look in, let alone prevail, sometimes in spite of Norm's best efforts. In short, we have brought together as many oppositions around risk as we felt we could in one book, and in the process we hope to start—or restart—reflection on that mighty clash of worldviews. There's also something or other going on in a subplot about asteroids and the end of the world.

We've already stated our own conclusion about how all this ends. But what that conclusion means, whether the argument stands, and how or whether the two sides can be reconciled is ultimately, of course, for you to judge, as you also explore where you stand on danger, if you dare.

THE BEGINNING

Had he not poured gin in the fish tank in a stupid reflex flick of the wrist when he tasted it—because the thing is, he hated gin—the whole story would never even have begun.

It wouldn't have happened if the fish hadn't died, either. Or if he hadn't felt bad for crashing the party in the first place and realized that the girl knew who he was and might tell on him. But for all that, he probably never would have been standing there the next day saying he was sorry.

"Erm, the fish . . . ," he said.

"Yeah, the fish," she said.

"Dead?"

"Yeah."

"Right. I kinda knew that. I think it was me."

"Uh huh."

"So, erm, how much is a fish?"

"For one fish? . . . Dinner."

"What? . . . Dinner? . . . Oh, okay! Yeah, dinner. But like, not like . . . not for every fish?"

"Hey, come on, fish killer!"

"Okay, Okay . . . "

"Though, to be honest . . . they weren't my fish. But it's either dinner or I tell my brother it was you and as he's a psychotic axe murderer, you don't want that."

"Right. Sure. So . . . how many fish?"

"Forty-two."

"Forty? . . .!"

And had they not then discovered a shared love of sudoku, sailing, and an original sound recording of Alfred Tennyson's "Charge of the Light Brigade," plus that he really liked her smile and she liked his hands and had a fascination for the strange birthmark on his right ear, well . . .

"Incredible," they often said afterward.

"What are the odds against meeting like that?"

"But 'forty-two'! It wasn't even true."

"Exactly!"

So it was only after a heap of happenstance, a whole cocktail of accidental, it-could-all-so-easily-have-been-different events, that they met again and talked and fell in love and had a baby—after he forgot the contraception the time they went camping but said "What the hell" anyway—which really should have made the whole saga about a zillion to one against.

But then, when you think about it, everyone is improbable, everyone's story's a fluke. There are so many reasons why any one of us might not have happened. At least, every particular someone is improbable. There'll be people for sure, but why you?

As it was, by going back to say he was sorry, he was out when his apartment caught fire and filled with suffocating fumes.

So as she lay screaming for an epidural and swearing that he was going to pay for this with his ass, was going to sleep with the fishes in fact, he was wondering about the baby's future, the strange course of luck and bad luck, the risks and coincidences of life, wondering how much in the riot of fates was calculable. What *are* the chances?

And at the very moment the baby was born, far away a spectacular fireball lit up the predawn sky, a radiant explosion caused by the atmospheric entry of a small near-Earth asteroid just a few meters in diameter but weighing 80 tons and firing icily through 12 kilometers of space per second, such that it shattered with the force of 1,000 tons of dynamite and the brightness of a full moon into small meteorite fragments across the Nubian desert below.[1] The asteroid was named Almahata Sitta. The baby weighed precisely 3,400 grams.* They named him . . . Norm.

. . . .

CAN NUMBERS HELP the infant Norm duck the slings and arrows of life? In *The Norm Chronicles* we will guide him with the best statistics that we can find. We will also make them as clear as possible.

The last point—clarity—is a big one. There are lies and damned lies in risk statistics, for sure, but there's real information, too, and a large part of the problem is cutting through to the good stuff and making it intelligible.

Say that Norm's dad is cooking sausages for the boy's dinner when his ears prick up to a headline on the TV news that says eating an extra sausage—or is it a sausage every day?—something about a sausage anyway—increases our risk of cancer by 20 percent. He pauses. Norm plays. The sausages sizzle.†

What does it mean, this percentage—20 percent more risky than what? Then he hears it referred to on the radio as a probability (and is that the same as a percentage?), and maybe later he'll read in the newspaper about the "absolute risk" and a "relative risk," and by now he's struggling, and who can blame him? But then there's another thing they talk about, called a "risk ratio," and it all seems mathematical and maybe even rigorous—who knows?—and so some say "you mean 20 percent of people die from sausages?" and others "you mean extra sausages cause 20 percent of cancers?" or perhaps "you mean 20 percent of people have a 100 percent chance of cancer if they eat, er, 20 percent more sausages?" And some, who

---

* Exactly the median birthweight recorded for children born in England, according to the latest available data, equal to about 7 pounds, 8 ounces, the same as in the United States. Centers for Disease Control and Prevention (CDC), "Birth Data," 2010, www.cdc.gov/nchs/births.htm.

† For an explanation of what the 20 percent increase in a risk does mean and how it's calculated, see the discussion in Chapter 4, "Nothing."

love sausages, say that it's all lies and statistics, but perhaps a few say, "Oh my God, it's a sausage, stay away!" and they hardly believe it, but maybe they should believe it, and some people tell them they're just stupid, but they don't feel stupid, they feel fed up, and still they haven't much idea what it really means, and in the end they say, "Oh, who cares, let's have another sausage."

Forget all this. We can do better. Norm faces a life full of such fears, often conveyed with about as much clarity. Can we ever calculate his precise fate? No. Obviously. No one knows the future. But as Norm grows up he can learn about the recent past—as with the body count for heart disease—and then extrapolate the average risk into the future and use this as a guide for his own life. This sounds imperfect but reasonable. In practice, risk, in the telling, is often a mess. Yet, so far as the basic body count goes and what it means to our everyday lives, it could all be easier to understand. That, at least, is one hope for this book, and for Norm, too—although danger isn't only something people fear and avoid, and maybe even sausages taste better if you think they're on the wild side. Either way, whether you seek danger or avoid it, we will try to make the numbers simple.

Our main technique will be a cunning little device called— by someone[2] with a wicked sense of humor—a MicroMort (MM), which is a one-in-a-million chance of death. MicroMorts are cheery little units that help us see danger in terms of daily life. They are risks reduced to a micro or daily rate on a consistent scale. The idea starts with exactly that: an ordinary day in the life of any average person, like Norm.

How risky is it for Norm, or for you maybe, to get up, go about your daily routine, do nothing particularly dangerous—no wing-suit flying or front-line duty in Afghanistan, just the ordinary—then come home and go to sleep? Not very, as you would expect.

True, you might lose your life under a bus rather than just your ice cream, slip fatally in the bathtub, or be murdered in a mistaken gangland revenge attack with a power tool, but it's not likely. It is a pretty micro risk. We know this. In fact, we can count the bodies and put a number on it. Typically, about 50 people in England and Wales die accidentally or violently each day by what are known as external causes.[3*] Since there are roughly 50 million people in England and Wales, this means that about one in a million gets it this way, every day. Not a lot, as we said. And even though you don't know for sure if you will be one of today's 50 or so who die from external causes, you're probably not lying awake worrying about it too much.

So this daily risk is around 1 MicroMort, a one-in-a-million chance of something horribly and fatally dramatic happening to Mr. or Ms. Average on an average day spent doing their average, everyday stuff—in the United Kingdom. In the United States it's a bit higher—roughly 1.6 MicroMorts a day, or 1.3 excluding suicide, or 1 if you exclude homicides as well to just leave the effect of accidents.[4] This, in other words, is a benchmark for living normally. You have experienced this, often. What's more, you survived. Congratulations. There you go, one-and-a-bit MicroMorts, today, tomorrow, every time.

Of course, this is just an average, and who except Norm is average? Some people are too timid to step out of the front door, while their neighbor revs the motorcycle on the way to a base-jump. What Your risk is (that's You reading now), is a much trickier question that we will come to later. For the moment, please imagine Yourself to be average.

---

* More precise figures for the United States: according to the National Vital Statistics Report, around 180,000 people died from "injuries" in the United States in 2010. That is all those people—out of the total population of about 300 million—who died from accidents, murders, suicides, and so on. This corresponds to an average of 180,000/300 = 600 MicroMorts per year for each person, or about 1.6 a day. If we take out the 38,000 suicides, it's 1.3. It is not a perfect benchmark, not least because of the question of how to treat suicide, which is not quite an "injury" but is categorized as one. But it gives a reasonable approximation of the hazards of daily living, and—provided we keep everything on this same scale—it is all reasonably comparable. Sherry L. Murphy, Jiaquan Xu, and Kenneth D. Kochanek, "Deaths: Final Data for 2010," National Vital Statistics Reports, vol. 61, no 4 (Hyattsville, MD: National Center for Health Statistics, 2013).

We will also have to assume that recent data give a reasonable idea of the future. In other words, we will move smoothly between historical *rates*—How many people died out of each 1 million?—and future average *risks*—What is the average number of MicroMorts per person? But by "future," we only mean the next few years—who knows what will happen after that?

With MicroMorts at the ready, the rest becomes relatively easy. How will you spend your daily MicroMort? If you ride a bicycle in the United Kingdom for 25 miles, that's your daily ration. Or you can achieve the same by driving 250 miles in the United States, also equal to 1 MM (remember this is an average—1 MM goes further on freeways). Or you can take a few more risks and add to your daily MM dose.

The joy of the MicroMort—if joy is the word—is that it makes all kinds of risks comparable on the same simple scale. And there are plenty of risks about. For instance, have you in the past ever been born? Do you expect to give birth? Do you now or will you ever drive or fly? Take drugs, including alcohol or painkillers? Ride a horse or bike? Climb Everest? Work down a mine? Climb a ladder? Spend a night in the hospital? Have you or your children ever had a vaccination? Do your toddlers put small plastic toys in their mouths in flagrant violation of the clear written warning on the packet? What's the risk that an asteroid is right on target for you?

All of these, and every other acute risk, can be measured in MicroMorts. For example, the risk of death from a general anesthetic on a "healthy patient" in the United States is roughly 1 in 200,000, according to the American Society of Anesthesiologists (although we might wonder why they are operating on healthy patients), meaning that in every 200,000 operations, someone dies from the anesthetic alone.[5] This risk is not as intuitively easy to grasp or compare as it could be. But we can convert it into 5 MMs, or around 5 times the ordinary average risk of getting through the day without a violent or accidental death, or around 20 miles on a motorcycle in the United States. We can show you, for example, that every shift working down a mine not so long ago in the United Kingdom carried, on average, about the same risk as going skydiving once today: about 7 MMs in the United States. A day skiing? An extra 1 MM, the same as an extra average day of nothing much. Anna would be reassured.

But at the extremes, 1 MicroMort is also the average risk incurred by US personnel every hour serving in the armed forces in Afghanistan in a bad period, making their service around 20 times more dangerous than average everyday living. Or it is the risk incurred by the aircrew of a World War II Royal Air Force bombing mission over Germany in around one second.*

A MicroMort can also be compared to a form of imaginary Russian roulette in which 20 coins are thrown in the air: if they all come down heads, the subject is executed.† That is about the same odds as the one-in-a-million chance that we describe as the average everyday dose of acute fatal risk.

While we're on the subject, take a moment to decide whether you would be willing to play this game of being executed if 20 coins all come up heads and if we paid you, say, $7 a go.

You wouldn't? $7 is not enough, you say. So how much money

---

* In 2010, out of an average deployment of 63,500 in Afghanistan, 499 US military personnel were killed, representing an average of 22 MicroMorts a day, a somewhat higher risk than in the year before (17 MMs per day), or in 2007 in Iraq (also 17 MMs per day). This rate, of course, averages over all personnel—those out on patrol would face hugely greater risks than those at a base. Between May and October 2009, out of 9,000 UK service personnel in Afghanistan, 60 were killed. This works out to around an average of 47 MMs per day, or 1 MM per half-hour. Between 1939 and 1945, 55,000 RAF bomber crew members were killed in 364,000 missions. With an average crew size of around 6, that's around 25,000 MMs per mission, or roughly 1 MM per second. Amy Belasco, "Troop Levels in the Afghan and Iraq Wars, FY2001–FY2012: Cost and Other Potential Issues," July 2, 2009, Congressional Research Service Reports on General National Security Topics, available at www.fas.org/sgp/crs/natsec/, cited November 18, 2013; "Operation Iraqi Freedom" and "Operation Enduring Freedom/Afghanistan," 2013, iCasualties, http://icasualties.org/, cited November 18, 2013; S. Bird and C. Fairweather, "Recent Military Fatalities in Afghanistan (and Iraq) by Cause and Nationality," 2010, www.mrc-bsu.cam.ac.uk/Publications/PDFs/PERIOD_9_10_fatalities_in_Afghanistan_and_Iraq.pdf.

† The chance of this happening is a half (heads or tails) times a half, repeated 20 times (1 in 2^{20}), which is roughly equal to one in a million.

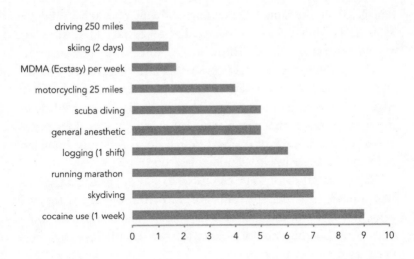

FIGURE 1. Selected risks in the United States in MicroMorts (1 MicroMort = 1 in 1 million chance of death).

would you want in return for accepting a one-in-a-million risk to your life? In other words, how much is your life, or a one-in-a-million threat to it, worth?

We can form an idea of how much governments are prepared to pay to save you from a MicroMort by looking at the Value of a Statistical Life (known in the trade as a VSL). This is a real-world concept used by governments to determine, for example, which traffic improvements to make. If a new junction is expected to save one life, then the US Department of Transportation might be willing to pay up to $6.2 million for it (in the United Kingdom an anonymous life is only priced at £1.6 million, around US$2.6 million).[6]

Therefore, the US government prices a MicroMort at one-millionth of this, or $6. Were you willing to play the Russian roulette game for $7? No? The government thinks you're overstating your value.

We will present risks in a variety of ways, but we will use Micro-Morts often. The MicroMort describes acute risks, those that hit you over the head with a "Thank you and goodnight." Later, we will introduce another measure, the MicroLife, to talk about longer-term hazards: chronic risks of the kind that slip slowly into the bloodstream and build up over a lifetime, such as cigarettes, diet, and drink.

Both measures take a few liberties and make some sacrifice of precision in exchange for ease of use as a daily measure of the hazards of life: one number, easy to compare with every other, rather than a muddle of percentages. Sometimes we will show the calculations involved, but we will tend to put these in the notes for those who want to skip them.

It is partly on the sense of proportion from MicroMorts and MicroLives that Norm will eventually stake his hopes for a fulfilled life, in which knowledge and comparison will guide him, help him realize his full potential to become the flourishing, uninhibited Norm he was always meant to be. So far, so reasonable—if reason is what counts. In time, we'll see. First, there's someone else we'd like you to meet.

2

INFANCY

From the day Prudence was born, her mother was never not afraid.*
She became protector, haven, she-wolf. She sniffed danger like a forest animal. Other children were germy, snotty, clumsy, other. Fungus and barbed wire, on legs.

Watch her now in the restroom, with thighs of iron, paisley skirt hoisted to her armpits, hovering. See how vigorously she then washes her hands. Meanwhile, notice Prudence in the stroller reach out with infant curiosity to touch . . .

"No, no, no! Dirty, Prudence. Don't . . . ever . . . Over here . . . "

"Mama."

"Hands . . . "

"Huh . . . "

"Oh you haven't? You have. All right. Legs up. Dipee . . . don't touch. Whoops. Where are the wipes?"

Wipes. Trusted companions. First line of defense in the war with dirt. Prudence would learn from an early age never, *never* to take peanut-butter sandwiches onto school premises, *never* to mix

---

* "Once she was born I was never not afraid. I was afraid of swimming pools, high-tension wires, lye under the sink, aspirin in the medicine cabinet, The Broken Man himself. I was afraid of rattlesnakes, riptides, landslides, strangers who appeared at the door, unexplained fevers, elevators without operators and empty hotel corridors. The source of the fear was obvious: it was the harm that could come to her." Joan Didion, *Blue Nights* (New York: Knopf, 2011).

the knives used for raw chicken. Just as her mother worried for the unaware, pitied them their lack of foresight and the risks they ran. Invited to join a book club, she nailed instantly the danger others missed—that if she lay awake to read at night, her husband might think she had the energy for sex.

The headline in her morning paper, "20 Physical Signs That You Are Seriously Ill," meant Prudence's Weetabix must wait while she read it, twice, and self-diagnosed. Was that a new mole on Prudence's leg? No, a spot of mud. Where were the wipes?

Follow mother and daughter now as they emerge from the restroom into the coffee shop, where an old friend sneezes. See how the mother leans back from the table, lips zipped, breathing as if air hurts, trying to disguise the sweep of her hand over Pru's mouth and nose. She's thinking of disinfectant hand-wipes.

"You need perspective," a male friend once told her, and offered to show her some probabilities.

"Thank you," she said, "but numbers aren't the point."

"What?"

Bad things happened in the world, unspeakable things. The thought of "what if?"—what if the worst happened—beat the numbers game every time. "What if" turned any risk, however small, into Prudence . . . screaming.

"What if I show you it's one in a million?" he said.

"No good."

"What?"

"The problem's the one."

"One in a million is good."

"Not to the one. Not to the one."

Especially if the one was Prudence. Was her childhood home child-friendly? It was. Including the garden ponds or water features? They were dug up. Did her mother know how to react in an emergency? She learned. Had she chosen safe equipment and furniture? She had. There was safe and healthy eating, safety on holiday, sleep safety, bath safety, safety on stairs and sofas, little objects and little bones (easily broken), scalds and burns, suffocation and drowning. Parenting magazines kept her alert. "Neglect" was almost the worst word she knew.

She also knew that some terrible twist of nature, an undiscovered illness, say, might still take Prudence. But short of that, and with proper care, she would grow up healthy, to outlive the reckless and misguided who drove too fast, ate badly, ignorant of cholesterol and acrylamide, grew fat, and, careless of gender-bending pollutants out there, left electromagnetic devices switched on in bedrooms and even had their children vaccinated. Or was it that they didn't have their children vaccinated? Bugger, what was the latest? And Prudence was due a shot. Sometimes the threat seemed to come from both sides at once, like two-way traffic.

Children on a winter's day meant a slip hazard; summer meant wasps and sunburn, with an eye for jellyfish on the beach and tsunamis on the horizon, when even restful moments with the *Mail* and a mint tea required a watchful eye over Pru's play, which certainly ought not to stray beyond the line of the deckchairs. An advertisement on the train, "Don't risk bad breath!" was on an all-embracing continuum with terrorists: "In these times of heightened security, please ensure you keep all your belongings with you. If you see a suspect bag or package . . . "

This was the world Prudence would inherit, a world of hazards, threats, risks, or symptoms in which numbers hardly mattered but a scary story might, in which fear was the price she paid for love, in which Prudence, hurt, was an infinity of pain and guilt for her mother, too—as she felt again on the way home after accidentally banging her daughter's head on the car door.

. . . .

PRUDENCE'S MOTHER IS RIGHT to this extent: being a baby is high-risk—relatively. The average hazard in the first tender year of life in the United States—around 6 per 1,000 births—is roughly that of riding 25,000 miles on a motorcycle, once around the world.* Imagine your baby on a Harley. Feel safe?

Those who survive won't face the same level of annual hazard again until their mid-fifties. Under-one is a dangerous age, comparatively.

But you probably knew that. Children's vulnerability is a cliché. First helpless, then a dicey mix of naive and curious, they're asking for it, aren't they? It's easy to see them all on the same high wire until they grow up.

Easy, but misleading. This splurge of fear is too crude. It masks a stark difference in risks that change radically in only a few years. Prudence's first year is relatively risky all right, but most of the risk is squashed into the first few weeks, and if she makes it to her first birthday, as the vast majority do, her annual risk plummets—from 6,100 MMs in the first year in the United States to fewer than 100 MMs a year for a ten-year-old, or about a quarter of a MicroMort a day from all causes. Believe it or not, this makes ten the safest age of all to be alive in the United States—far safer than life for mom and dad.[1]

So infancy and childhood go in short order from one extreme of life's acute risk to another. Babies don't do much, but they do live dangerously, briefly, compared with others. As soon as they become young children, they become more robust and are safer from death than anyone.

This is only fatal risk, and there's plenty of damage you can experience short of the big one. Even so, the change in fortune is stark. Anxious parents have something to worry about in year one, most of it in the first weeks, and then . . .

But does their anxiety fall with the risk? Or do they worry away anyway, like Prudence's mother? If so, maybe it's watchfulness that keeps children safe, in which case worry works. Certainly, she thinks so. A danger foreseen is half avoided, an old proverb says. Or

---

* Assuming it's all on US roads. The risks on an Afghan pass are probably higher.

maybe once parents start to worry, they can't stop, and it takes about a fortnight for paranoia to set in and never let go.

But there's another reason Prudence's mother is frightened: not just love but also fear of blame. Danger can feel worse if you know that the outside world will look for fault when things go wrong, if you know that people will want to name the person who "did it." Then there's an innocent victim, and an accident turns into something more damning than bad luck—it becomes a grievance. If you are the one doing the looking after, when protection is your duty as well as your care, if you are not just loving but also the one who'll be blamed, then danger is darker—and guiltier. "And where was the mother? . . ." someone will ask.

So if something happened to Prudence, how would her mother feel? Amid everything else, rightly or wrongly, she'd feel she hadn't been there. You could tell her it was bad luck all you liked. She wouldn't hear. Probabilities don't mention this—the human pain.

But don't judge them (the probabilities, that is) too harshly. This indifference to people's feelings can be a virtue. For not all the feeling in stories is benign. It can be bitter, resentful, eager for revenge. And so probabilities, which aren't interested in any of this, can be kinder than stories, more forgiving, less eager to find human agency (a particular person who "did it"), and happier to admit their uncertainty about exactly how the accident happened. So probability is uncaring, it's true, but by the same token it is often un-blaming.

That's partly what probability is—a statement of uncertainty about cause and effect or naming the guilty. Being an emotionally challenged number can have its humane side, too, if only by omission.

What the numbers primarily say is that the most serious risk of death in infancy is brief, as we say, and also that it has plunged. Infant mortality is a good indicator of the change in social conditions, and strongly influenced by transient famine and epidemic. It tells its own story of breathtaking progress.

Throughout ancient history, perhaps 30 to 40 percent of babies would die before their first birthday, what we might call the natural rate of infant mortality. By about 1600 in England, the figure had roughly halved, and it remained at about 15 percent from the mid-1800s.[2] If this were the rate today, it would mean well above 100,000 infant deaths a year in the United Kingdom, or 600,000 in the United States.

Fortunately, it improved again, dramatically. By 1921 the rate had almost halved for the second time, but this time the leap took just one generation. Soon after World War II, it had halved again. By 1983, even that rate had halved again and then halved again, and then by 2012 it more than halved again to stand now at about 4 per 1,000.* The change has been nothing short of astonishing, death slashed back, again and again. In much of the world, the biggest historical cause of early death has been so reduced that what was once ordinary is now exceptional. Has there ever been a more radical reduction of risk in human history than the risk of infant mortality in developed countries?

How do those numbers make you feel? Privileged by progress and reassured for your children? Or still anxious?

For though the risks have reduced massively, they are still around. By 2010 in the United States there were 4 million live births a year, more than 8 a minute, or close to 11,000 a day: enough to fill 20 very noisy 747s. Think of Prudence as one of them. So far, so good. But she had to survive several scary moments along the way.

The first came before the first breath. If you look around that crowd, you notice 70 empty seats. These represent the stubborn toll of stillbirths, around 1 for every 160 live births, a rate unchanged since the early 1980s. While other countries have continued to make progress, both the United States and the United Kingdom have not. This is a puzzle, and it is disturbing.

Next comes survival in life. Of the 11,000 live babies born each day, around 35 die in the first week, another 9 before a month is out, and 22 more before they reach their first birthday.[3] The total

---

* Although infant mortality had fallen sharply up to 1921, it still accounted for a staggering 82 babies out of every 1,000 born in 1921. Then it dropped sharply in only a generation, to 46 in 1945. By 1983 it was down to 10, and it is about 4 now.

risk during that first year from a variety of causes is, as we say, 6,100 MicroMorts. That is how we arrive at the 25,000-miles-on-a-motorcycle equivalent. If 1 MM equals riding a motorcycle for about 4 miles, then 6,100 MMs × 4 miles = 25,000 miles.

We'll take the dangers to Prudence that make up this total one at a time. First, the largest categories: congenital disease or prematurity, which account for over one-third of infant deaths.

As Prudence was not premature or congenitally ill, the 25,000-mile motorcycle equivalent risk falls sharply, to about 15,000 miles, from 6,100 MMs to about 3,800 MMs.

But then comes the next danger: 2,500 babies, 8 a day, who died after complications of pregnancy or birth. In the United Kingdom it is a continuing controversy whether it's safer to have your baby at home or in a hospital (at least, since hospitals learned the lessons of hygiene; see Chapter 11, on giving birth). People who give birth at home are generally better off financially, and that tends to be associated with higher rates of survival. They also tend not to be having their first baby, which should also be safer. Even so, the infant mortality rate is the same for home births and hospital births in England and Wales.

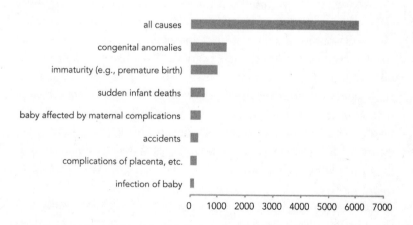

FIGURE 2. Causes of death in infancy in the United States, approximate risks in MicroMorts, 2010.

A recent study of 65,000 "low-risk" births showed that women giving birth in a unit run by midwives were just as safe as those in a traditional hospital, but had far fewer Caesarean sections and more "normal births." If it is not your first child, then giving birth at home is as safe as in a hospital, but for first-time mothers there was around double the risk of a serious problem at home, and nearly half had to be transferred to a hospital.[4]

The final big danger to Prudence—and everyone else—is sudden infant death syndrome (SIDS), with 2,063 cases in the United States in 2010 (500 MMs). In the United Kingdom, since the launch of the Reduce the Risk campaign in 1991, which overturned the received wisdom of leaving babies to sleep on their front, the number has dropped by 70 percent.[5]

These mysterious deaths are around 50 percent more common in boys than in girls and are more common in winter than in other seasons of the year. The risk is 5 times higher for children of mothers under 20 than for children of mothers over 30. So all in all, as a girl who was not premature and had a mother over 30, Prudence was among the safest at this—relatively—dangerous age. Not that her mother would have been reassured.

And if Prudence were born in the United States, she would be safer, on average, if she was born to a white mother—infant mortality is still double for babies of black mothers.

That is in a developed country. What if she had been born elsewhere? International comparisons for infant mortality are tricky. Some countries don't include in their infant mortality data the tiny premature babies that we have seen are at such high risk. Lack of good registration of child deaths also means that the rates have to be estimated from surveys that ask households if they have had a child under five who has died, then by using statistical models to estimate infant mortality.

The admirable UN Inter-Agency Group for Child Mortality Estimation puts all these data together and estimates that, over the whole world, the average risk faced by a baby in the first year is around 35 per 1,000 (a chilling 35,000 MMs), about the level in England and Wales in 1948.[6]

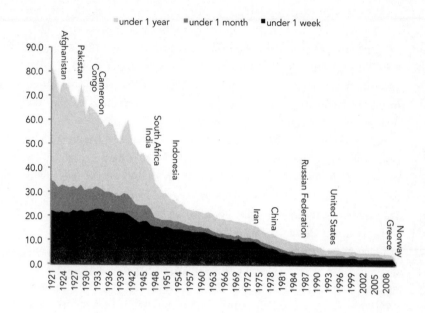

FIGURE 3. Historical trend for infant mortality (death before 1 year of age) in the United Kingdom, with the position of the United States and selected countries, 2011.

But, like so many averages, this one obscures massive variation. At the bottom are Sierra Leone and the Democratic Republic of Congo, with rates of around 117 and 100 per 1,000, respectively, the English rate in about 1919. Ethiopia comes in at 47 (the English rate in about 1945), Vietnam 18 (1969), and the United States, as mentioned above, at 6 (1997). Cuba's 4 per 1,000 is similar to England, while Finland and Singapore are down to 2 per 1,000, about half the rate of the United Kingdom.[7] This suggests that we cannot separate the world into "us" and "them"—"they" are mostly like us a generation or so ago, and catching up fast.*

The line in Figure 3 shows the dramatic improvement in the United Kingdom's infant mortality rate since 1921, measured in deaths per 1,000 live births before one year old. We've also shown

---

* See Hans Rosling's Gapminder project (www.gapminder.org) for brilliant demonstrations of this principle.

the recent position of a selection of other countries. Thus Afghanistan in 2011 was about where the United Kingdom was in the early 1920s.

Infant mortality in the United States has followed the same pattern, falling from 29 per 1,000 in 1950 to about 6 per 1,000 in 2011, an admirable drop of 80 percent. But the rate for whites has fallen more (81 percent) than the rate for blacks (74 percent), making the relative gap between these groups even larger.[8]

The Millennium Development Goals were set by the United Nations in 2000. The fourth goal was to reduce infant mortality by two-thirds between 1990 and 2015, from 61 per 1,000 live births to 20. The current level of 40 shows great progress over twenty years, but means the overall goal is unlikely to be met, although some countries have made giant strides: for example, Malawi has gone from 131 to 58 per 1,000, and Madagascar from 97 to 43, both of which are 56 percent reductions in twenty years.[9] If you have the impression that the developing world makes little progress in human welfare, think again. The number of babies who died before their first birthday is estimated to have fallen from 8.4 million in 1990 to 5.4 million in 2010—an astonishing improvement, but still representing 15,000 a day, 600 an hour, 10 a minute, 1 every six seconds.

Here's an odd question: Is the death of a baby sometimes just nature's way, as Prudence's mother fears it might yet be, even after she has done her best to eliminate all other risks? The distinction between what we're willing to call natural, and what we're not, matters. Many people feel that unnatural risks—risks associated with modern living, such as travel, technology, or obesity—are worse. Humans created these risks because enough of us liked the benefits from cars, nuclear power, or cakes. But if what we create also messes with nature, well, some say, what do we expect except our comeuppance? There's more than a hint of blame involved here. Unnatural risks sound like risks that somebody caused. So they're not just unfortunate, they might also make us angry. Maybe that's why it feels, to some, that it would be better to go by a lightning strike than a downed power line, as one piece of research put it.[10] Maybe these unnatural risks feel worse because this is what happens when humans get above themselves, or maybe they feel worse because someone else imposed them on us. In contrast, we might call the death of a very small child more like supreme bad luck, an act of God or nature, primitive or given, especially when illness is the cause.

"Natural" risks are often more tolerated. But it's an awkward attitude, and this chapter tests it to the limit. For what greater fear is there than that of seeing your child die? The narrative is so disturbing that even TV uses it sparingly, while portraying adult death on an industrial scale.

The first, obvious problem with this attitude is that disease is natural and slaughters people in the millions. Should we tolerate it just because it's always been around? The way that we apportion blame doesn't seem to be consistent. And yet "unnatural" remains a potent criticism, even though humans brought a dramatic decline to the natural rate of infant mortality, sometimes with nothing more unnatural than germ theory and soap. Few people think we should not have messed with it.

Although the anti-nature argument can also be taken too far: Does it mean technology is always good? Obviously not, even for children's health. Neither side wins every time. The natural/unnatural risk argument rumbles on around home birth (and vaccination—see Chapter 6).

So is this odd preference for natural risks irrational, when natural risks can be the most devastating of all? Strangely, maybe not. "Unnatural" can be a way of saying that you don't like a risk for other reasons. Maybe what you mean is that an unnatural risk is when people, governments, or big businesses are up to no good. Maybe you think someone is making too much money or seeking too much power. "Natural" and "unnatural" might be bound up with tricky ethical or moral feelings about the way a whole society behaves, and have nothing to do with hard and fast distinctions in nature itself, whatever that might mean. This doesn't make these feelings irrational, but it does make them complicated. Risk is often like that: ostensibly about danger, but really packing a whole lot of attitude about a whole lot more.

From thinking of one child, our own, and trying to imagine the loss, to thinking about the same loss multiplied more than 5 million times every year, then of another 3 million who now survive thanks to economic development, technological progress, and the medical control of disease, seems to leave little room for sentiment about risk and nature. But the sentiment persists. What does that say about us?

That risk, even a risk like this, is a small part of a sprawling human calculation about what's right and how to make progress, a calculation that's sometimes political, sometimes moral, and sometimes human vanity.

VIOLENCE

Harry the hawk's gimlet eyes policed the city's criss-cross streets on behalf of pest control division—"Rat-Swat"—of the municipal department of environmental health. High above, he twitched his feathered wings and rode the air, watching.

People, about their business. Vehicles, moving and stopping. Trees in the breeze. Children in danger.

There by the park, for instance, walked nerdy Phil, now in long trousers, not far from the older guys hanging out in the park. Phil stopped to stare at a puddle. The puddle fizzed. From beneath the broken words "Central Electricity Board" on a half-sunk and vandalized iron cover came pops and flashes.

"Idiots," said Phil, a tad too loudly as he stepped around the water. "They've actually electrified the pavement!" Harry saw the guys walk over. There was shouting, a scuffle, a knife. Phil fell—with a splash.

Across the street, Harry could see Mikey leaning on the railings. Mikey had been there a good fifteen minutes, had made the call on his Blackberry to report the puddle shortly before a stranger's hand scooped up the small backpack with his laptop, school textbooks, and notes that he had left leaning against the mailbox.

"Why is he wriggling in a puddle?" said Norm, who was nearly three, as he stopped beside Mikey, who had turned away from the railings and was now crouched and hiding behind the mailbox, tapping the phone again and not answering him and saying a word Norm thought might not be a very nice word.

"Are you hiding?" Norm said.

He was a bit frightened before, when daddy went and didn't come back somewhere and told him to "stay here," but he couldn't remember where "here" was. Now it was cold. And the man with the phone looked at him in a funny way. Norm wondered if he was a stranger.

"Because we don't want him getting a chill," said Mrs. Assabian, just down from the puddle at number 38, as she wrapped Artemis in his fleece jacket and tightened the buckles under his little tummy. "Now go," she said to her nine-year-old daughter Jemima, whose other bruises had largely healed, "or you'll get another," and clipped on the lead while shoving them down the steps, turning back inside and closing the door.

Harry registered in fine detail the movement of these two small figures, one upright, one horizontal. Horizontal had four legs. This was important. It scuttled from left to right, intermittently jerking forward, a pattern Harry recognized by instinct, honed by reward. He watched. Then he swooped.

Bloody big rodent, thought Harry, as he sank his talons into the red coat with the strange markings and began to haul the vermin aloft. But the vermin was stuck. Harry flapped harder.

Mrs. Assabian's daughter, by now on High Street at the window of a shoe shop, felt her right arm wrenched to the left behind her then yanked into the air. Twisting, she met Artemis at eye level, his paws dangling, snared in the claws of a beast beating the air.

"Noooo!" she screamed, "Get off, get off!"

She pulled and screeched, fighting as she had learned to fight other blows. Harry screeched too, louder, and flapped harder. A male passerby spotting a once-in-a-lifetime chance of tug-of-war for an airborne chihuahua, joined in, laid a hand ineffectually on the leash, and absolutely screeched.

Shoppers turned—and screeched. Artemis screeched, as best three pounds of dog can, and then, with a rip, fell to earth. Harry flapped away toward the river.

Where the crescendo of screeches carries the short distance over the city to find the four-year-old Prudence in the car seat as her mother drives them home from a trip to the big city. The traffic noise surges around them. Mommy beeps the horn. From somewhere behind them, another horn blares, livid and too long.

"What's that, mommy?"

"Silly man."

A small van behind them whines up through its gears. The van is veering and tilting insanely across the lanes. It squeals and stops in front of them. They brake hard, inches shy of a crash, and the horn blares again.

Blind to the traffic that checks and veers around his van, Van flings open the door and jumps out, engine running. Prudence's mom tries to pull away, but her car lurches and stalls. Head down, she twists at the key, pulls the wheel hard left, lurches, stalls again, twists, starts, revs high in panic.

Too late. Van has swept around the front of his van, and as Prudence's mom begins to drive, he leaps at the side of the car, clinging to the rim of the roof with one hand, feet unseen wherever they can hold.

Tongue out and leering, but silent, Van's face is pressed to the driver's-side window, his other hand a fist, beating at glass, door, windshield. She accelerates. Van is still there, a limpet, an alien, a beating fist. He has to jump. He must. He holds on. Still he holds on, then jumps backward—and spins onto the road.

It has taken just seconds. Not a word, not a shout, plumes of breath in the cold, Van saunters to his van, settles in as he moves off, reaches for the still-open door as it waves out against the turn of the van, pulls it shut, and drives off.

Prudence is silent. So is her mother, staring straight ahead as she grips the wheel. Never complain, never look, never speak out, keep away, strangers, mayhem, accident, danger, madness, hatred, death.

High above, Harry circles.

They drive home. In the column asking if the parent gives permission for the school day-trip, Prudence's mother circles, "I do not."

. . . .

FOR A PRICE, Norm's parents could have his genomic DNA isolated and quantified using agarase gel electrophoresis, in which "a region of gDNA is selectively amplified by polymerase chain reaction (PCR)" for a home DNA storage kit.

Why? In case he gets lost—as he does. You could do the same for your own children. At least, that's the least anxious motive. The darker thought, the thought of what DNA is used for, is to detect crime. And that means abduction, or worse. The sooner officers have these DNA details the better, say the instructions on the tin. You can also have your child's tooth-print recorded on an arch-shaped thermoplastic wafer. Let's not dwell on the moment that comes in handy. Fingerprints, identity bracelets, and recent photos are often recommended.

Abduction is a scary thought. Does this kit make it better? Would you feel happier for Norm knowing that we had his DNA in a tin? That's the problem with fear, a cruel, insidious thing: finding ways to alleviate the danger that provokes it means dwelling on the danger.

Could we ignore it instead? The chance that a stranger will try to abduct Norm or any other young child is on average extremely small; the chance that the stranger will succeed is far lower than that. The chance that someone will murder your child is smaller again. For murder by a stranger of a child under five, even that risk drops 97 percent. If you are a parent, by far the biggest violent risk to your child, on average, is you.

Even including the risk from parents, the risk of death from all causes for young children like Norm is the lowest it has ever been, and this is also, as we saw in the previous chapter, the age of lowest risk in an average human lifetime. That is, based on the most recent data, early childhood is the safest age to be alive, at probably its safest moment in history.

So what? Bad things happen, as Pru's mother says—and she is right. Our fiction is an absurd batch of shocking dangers and calamities—deliberately so—especially in the space of a few minutes and a half-mile radius. But they are all loosely based on real events. The electrified pavement and the curious incident of the hawk and the dog did happen, within a few hundred meters of one another. Every so often a child dies in an animal attack, though it tends to be dogs, not birds of prey. The road rage happened, too, though all the characters have been changed here and some events altered. Some may remember, more somberly, the story of Ryan Jones in 2007, an eleven-year-old shot dead in a pub parking lot in Liverpool, an innocent in the way of a gang feud. Or Baby P, as he was known, a seventeen-month-old boy who died after suffering more than fifty injuries over eight months at the hands of his mother, her boyfriend, and his brother.

These were horrifying cases. Our problem is that horrifying cases warp our sense of the odds. Precisely because they are rare, they stand out. They're more salient. One outrage now and then against a background of low odds stands out more than if the danger is common. It is a troubling trade-off: we pay for rarity with more acute sensitivity.

So is it despite, or because of, the fact that the chance of violence to young children is so remote that parents have grown—according to some writers—more anxious about their children's safety?* The salience paradox is one putative explanation for this split between low odds and high anxiety. The safer life becomes, maybe the more horrifying the extreme exceptions seem.†

So does this gap between the odds on disaster and how anxious we are prove we're paranoid and irrational, or that the data miss

---

* A growing fear for children has been dated to the 1960s, with the emergence of battered child syndrome, followed by runaways and Halloween sadism in the late 1960s; sexual abuse, child pornography, and kidnapping in the 1970s; and missing children and ritual satanic abuse in the 1980s. J. Best, *Threatened Children: Rhetoric and Concern About Child-Victims* (Chicago: University of Chicago Press, 1993).

† Although there is a problem with this "rare is worse" feeling. If it also suggests the obverse—that "more common is not so bad"—then it is not so far from the attitude that life must be cheap in developing countries: "So many people die, can it hurt much to lose another?" Readers can decide whether death feels worse if it is also more unlikely.

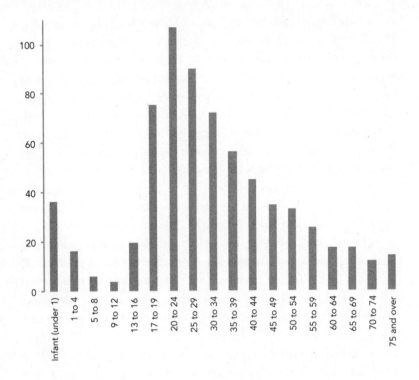

FIGURE 4. Annual risk of being a murder victim in the United States, by age, in MicroMorts.

the point, or both? To the rationalist who says, "But don't you realize how rare these things are?" there is a paradoxical answer: "Yes, which is why they are so alarming."

Here are some of those odds for American children.

The risk of murder is lower for children aged 9 to 12, at 4 MicroMorts per year, than at any other age. It's not much higher for 5- to 9-year-olds, either, at 6 MMs. On either side of this low point, the numbers rise. By ages 17 to 19, the risk has jumped sharply (79 MMs). But it's clear that whatever our fears, younger children are among the safest of any age. It takes a whole 12 months before they are exposed to a murder risk equivalent even to the normal hazard of acute death from nonnatural causes that an average adult is exposed to every few days.[1]

That's children. For infants, it's a different story. Those aged under 1 year old face a risk about 9 times higher than those who are 9 to 12 years old, at about 36 MMs a year. That's still low compared with the daily hazard of average living, being a tenth or less of the adult baseline everyday risk of death from all nonnatural causes, and still low compared with many other age groups.

But still, as Pru's mother says, it happens: and it did happen some 1,000 times in 2012 to people under 18 in the United States. A more homely, more human, and more sinister way of describing this is that about 20 children are murdered in the United States every week.

In the United Kingdom, the figure is about two child murders a week. So the United States has a child murder total about 10 times the United Kingdom's, but since it also has a population about 5 times bigger, the figures suggest that American children face roughly double the danger. The UK rate and MicroMort risk from murder for children is about half that of the United States.

Comparisons between murder rates in the United Kingdom and the United States can be controversial, and differences in definitions create some uncertainty around them. For example, in the United Kingdom, it is only murder when someone is convicted of murder. In the United States, if there's a body and officers suspect murder, then it's counted as murder. As the FBI states, "The classification of this offense is based solely on police investigation as opposed to the determination of a court, medical examiner, coroner, jury, or other judicial body."[2]

This definitional fog will make a difference, for example where the question of whether the US authorities have counted as murder a case that is subsequently determined by the courts to have been self-defense, for example. So it's hard to be sure that you are comparing apples with apples. The judgment you would have to make is how much this is likely to matter to the overall comparison. Our judgment in the case of child murders is that "double" is a lot to lose in the definitional margins. Some real difference in risk probably remains.

In other ways, the two countries' patterns of child murder have a lot in common. For instance, on the question of who does it. If you can rule out murdering your own child, the risk of your child who is under the age of 5 being murdered collapses by about 60 percent in the United States, and by a similar percentage in the United Kingdom. If your friends and other family members can be trusted, there goes another chunk. The people we really fear, strangers, are a small part of the total risk. In fact, parents are about 20 times more likely than strangers to be the murderer of a child (63 percent of child homicides were committed by parents, and just 3 percent by strangers).[3]

For abduction, the overall chances are much higher. According to what the National Center for Missing and Exploited Children (NCMEC) describes as the most recent comprehensive study of child abduction, every year something like 260,000 children are abducted in the United States. If we take the population of children under 18 years of age at the time of the study to be about 70 million in the United States, that's a rate of about 380 per 100,000 children.[4]

But again, the people parents should be most afraid of are their partners and relatives. Of these 260,000 abductions, about 200,000 were taken by relatives, often parents disputing custody. The rest were taken by nonrelatives with primarily sexual motives. The FBI says that despite media concentration on known sex offenders, these are also a small part of the problem, though a scary one, to be sure. In 2010, registered sex offenders were responsible for 1 percent of US child abduction cases.

At the most extreme end of the spectrum, about 115 cases were "stereotypical" kidnappings in which strangers abducted and killed children, held them for ransom, or took them with the intention to keep them. That's about 0.04 percent of all abductions.

But murder and abduction statistics count only the worst of what happened. And the difference between what does happen and what might happen is one of the ways in which risk calculations seem to some people to miss the point.

The figures count actual abductions or obvious attempts, not malicious intent or children's vulnerability. So let's try to capture the sense of potential risk, rather than only the real incidents, and count all the cases of missing children—as just one among many possible ways of measuring vulnerability.

Now the numbers explode. According to a study published in 2002, the number of officially reported missing children in the United States was about 800,000.[5] Again, a tiny fraction of these were "stereotypical stranger abductions." Among the rest were runaways and "throwaways" (abandoned children), children abducted by family members, children missing with a benign explanation, and lost or injured children, and this number also includes some children who went missing more than once.

This is a wide range of seriousness. Even so, excepting the benign explanations, as a British report on child abduction put it: "When a child goes missing, there is something wrong, often quite seriously, in that child's life. . . . Abuse, exploitation, and risk to life are the most concerning of all dangers that children face. Other risks include violence, criminality, and loss of potential due to lack of school attendance or other education, lack of economic wellbeing, sleeping rough, hunger, thirst, fear and loneliness."[6]

But why count only missing children? Why not all children everywhere for the measure of who might fall victim? Or is that the route to paranoia?

Part of our anxiety is over who might be out there with a vile motive. Until these people commit an offense, that is hard to know. Afterward, the authorities try to track offenders through registering sex offenders. Since some are tracked for many years, even for life, the total number of registered offenders has risen relentlessly since the tracking programs began. That does not necessarily imply a growing threat, though there is anxiety in both the United States and the United Kingdom that the Internet has created new incentives and opportunities for child abuse. But the rising number of people on the registers also reflects the fact that we have not been counting these offenders for long, and so it is partly just with passing time that the numbers are adding up.

In the United Kingdom, a child protection charity, the National Society for the Prevention of Cruelty to Children (NSPCC), says that of about 40,000 registered sex offenders, about 30,000 committed offenses against children.[7] In the United States, there are about 750,000 registered sex offenders. Does that mean there are about 19 times as many sex offenders in the United States as in the United Kingdom? Probably not. The difference is too big to be plausible. It means that's how many are registered, not how many offenders there are. The rules on who has to register vary from country to country; they also vary from state to state, making comparisons hazardous.

Some parents who know that the "stereotypical" crimes are rare are probably not too worried about the risk to their children. Some worry in crowds. Some can scarcely bear to let their children play outside. Some scan the sex offenders' register. Some are driven to fury or terror when a crime hits the news. In other words, we vary, but most of us have a threshold at which anxiety kicks in—and whatever that level is, we all feel that ours is the appropriate level.

Though some do say that this is all a "moral panic." The phrase was popularized in the 1970s in an analysis of the Mods and Rockers of the 1960s written by sociologist Stanley Cohen. In brief, the argument is that the media overreact to behavior that challenges social norms, and this overreaction comes to define the problem and even creates a model for others to copy.[8] It is a powerful analysis. But the word "panic" suggests irrationality, as if you are a parent, and you see another parent lose a child to a violent stranger, and your heart screams, but you must not pull your own children closer. Is that reasonable? The problem is that there is no tape measure—in MicroMorts or otherwise—for the shock of a child violently killed. Was the shock disproportionate in December 2012, when twenty children were murdered by a gunman at Sandy Hook Elementary School in Newtown, Connecticut? Does that shock go away if you know that there are roughly 15,000 homicides in the United States each year? Or does it grow? Who is to decide when proportionate reaction ends and overreaction starts? Probability alone isn't the answer.

4

NOTHING

The big saucepan on the stove with water in it made funny plopping noises. Steam hissed out. The egg inside rattled. The blue fire underneath was jumping. Prudence went to see.

Usually, mommy didn't let her—and turned the handle away. She'd be next to you if you did something naughty and she said it was dangerous.

"Prudence, are you there?"

She'd say, "No" (before the scalds, the shrieks, the inconsolable pain).

"Prudence? Don't wander off, love."

Today the little girl wanted to pick up the saucepan like mommy picked up saucepans, and she went to the gas and the bubbles, nearer, and she reached for the handle, and mommy came in, picked her up, and turned it off.

"There you are! And it's ready."

Meanwhile, six-year-old Norm was sulking.

"I can't find it!"

His father turned around in the driver's seat. The lights changed. The car behind them honked. He jolted back in his seat and glowered in the mirror, thought about sticking a raised finger through the window, as seemed to be the fashion nowadays, then thought better of it and pulled away. But Norm was convinced they'd left it behind.

"It's not here!"

His father turned again to reach back and rummage in the pile on the seat, then change a gear, turn again, glimpse, and accelerate.

You couldn't blame the driver of the truck coming in the other direction. He had no chance. He scarcely touched the brakes as Norm and his dad veered over. In one awful moment the cab of the 18-wheeler loomed and filled the windshield, and Norm's dad popped the wheel over to get out of the way, then pulled into the bus stop to look for Norm's sudoku properly.

"Be careful, Prudence, you might fall."

"But I might not."

At home, Kelvin lit a match and dropped it into the ashtray on the kitchen worktop. He watched as a wicked flame got the napkin, but then caught a corner of kitchen roll that was near, then the tea towel. The fire was growing. Kelvin stepped back. He didn't mean that. What had he done? Before his parents in the living room could separate the smell of their own cigarettes from real smoke, fire had taken the curtains and Kelvin was staring at a sheet of flame. His dad raced into the kitchen to find that pretty much everything flammable had gone up—there wasn't much—and the fire had petered out.

By evening, Prudence's mother stood in the doorway, arms folded, eyeing her husband sprawled on the sofa in front of NBC's Brian Williams. Useless waste of space he was, her best kitchen knife lying ruined on the coffee table where he'd left it after changing a fuse on a plug because he couldn't be bothered to go out to the garage for a screwdriver. Look at him. Useless, always in the way, with not an interesting word to say, and what could she have seen in him? And ten years of resentments poured through her. Thirty more of ghastly married solitude reared up. And she stared at the knife and felt a despair that flared into a rage so sudden that in one swift movement she strode over and took the knife and felt the strength of vengeance in her right arm for all her murdered hopes as she saw the terror in his pleading little eyes and raised her hand and plunged the knife deep into his chest in one of those trivial fantasies of resentment that came and went in a blink before she sighed and smiled, and went over to pat his head.

"Ready for bed, love?" she said.

. . . .

IN A SKETCH BY British comedians David Mitchell and Robert Webb, a filmmaker (Mitchell) is interviewed about his *oeuvre* after a clip from his latest work: *Sometimes Fires Go Out*. In the film, as a couple watch TV, a small fire in the kitchen—yep, you got it—goes out. That's it.

The interviewer (Webb) says the film has been reviewed as "unrelentingly real," "a devastatingly faithful rendition of how life is," and "dull, dull, unbearably dull"—all those comments, oddly, from the same review.

He introduces a clip from another film: *The Man Who Had a Cough and It's Just a Cough and He's Fine*.

Two Edwardian lovers have a series of rendezvous on a station platform. The man, spluttering, looks more pallid and doomed with each encounter. "It's just a cough," he says, stoically.

Except that it is. It's just a cough. In the last scene he's dandy. It is one of the finest comic sketches about probability you'll ever see. But then, where's the competition? The idea is lovingly ripped off in our fiction, with some added examples.

Explaining jokes is a bad idea, and the joke here is simple: stories are about what happens; they're not often about what doesn't. If nothing happens in a story, it is not usually a story, it's a joke. And that, funnily enough, is a problem.

Fictional stories choose what to bring to our attention, or their authors do, and whatever they choose, they choose for a reason, often to prime the reader for what comes next. Anton Chekhov said: "If in Act I you have a pistol hanging on the wall, then it must fire in the last act." Or imagine an episode of the hospital drama ER, the family gathered around the breakfast table, when the old man coughs . . .

You know what's coming next. A cough in everyday life is not what you might call a statistically significant event. But you're watching ER. And so you know that this will be a triple heart bypass.

And you are right. Likewise, a pan of boiling water and a toddler can mean only one thing in fiction, and it must involve screaming. In stories, do fires ever just go out?

Risk is much the same. It is almost always framed by the thought of events, not by non-events, framed by things that happen, not things that don't. As soon as we talk about the risk of heart disease, we are thinking of those who will die of heart disease, not those on the other side of the odds who will be fine. The whole discussion of risk is primed by the thought of the bad stuff that happens. As Chekhov should have said, everything you hear about risk begins with a revolver on the wall.

All this may be true, you say, but how else can we talk about risk? What's the alternative? The alternative is to approach the whole subject with our focus on what doesn't happen, the non-events, the coughs that are just coughs and the fires that go out. Then risk would no longer be only about the deadly, unhappy endings, it would also be about your chance of being fine if the risk didn't, in the event, burn down the house.

This is an unorthodox take on the subject, admittedly. You'll look in vain for the headline "No Children Killed on the Way to School Today." Non-events are by definition systematically neglected in news coverage. Imagine the newsroom conversation that begins: "Anything interesting not happening today?"

But is the concept so daft? As we said earlier, a probability like 30 percent implies a non-probability of 70 percent. There are two numbers, not one: two sides to the odds. But the way risk is usually framed puts more emphasis on one of these numbers than the other—that's what "risk" tends to be taken to mean—with no equivalent word for thinking of the numbers the other way round. What is the opposite of risk's emphasis on events? Not safety, exactly. That doesn't quite describe the nothingness we're after. The dictionary and thesaurus are not much use, offering a meager list of synonyms for non-events, such as "damp squib," which hints at how alien the concept is. We scarcely have words for it. But that's because it is unfamiliar, not because it's ridiculous. How would a newspaper describe the thing that didn't happen, if it ever did? Would it see the point?

Take a real example. "If you're the type who eats 150 grams of processed meat a day—from only three sausages or six strips of bacon—your pancreatic cancer risk shoots up 57 percent," CBS news reported online.[1] Not only does this way of stating the risk put the cancer revolver on the wall; when it also says that the risk "shoots up," it caresses the trigger.

How could we reframe this risk to emphasize the fact that in the real world nothing also sometimes happens? First, let's change the units a little to make the numbers nice and round. The same study referred to by CBS puts the risk for every 50 grams (about two strips of bacon) at about 20 percent (a third of the quantity of bacon, a third of the additional risk).

Next, let's note the emphasis in the original report on events. This "20 percent up" is a calculation based only on the things that happen. It begins with people who have or will have cancer and ignores all those who don't or won't.

The US National Cancer Institute says that about 1.5 percent of Americans currently develop this (very aggressive) cancer during their lifetime. That's about 5 in every 400. If we take the figures at face value, the 20 percent increase in risk if everyone—*every one of those 400 people*—eats two extra strips of bacon *every* day—and good luck to all who try—takes that toll from five cases to about six. That is how we calculate a rise of about 20 percent. Twenty percent of the original five is one additional case. We have gone from a "relative risk," a 20 percent change, to a change in "absolute risk" from about 5 in 400 to about 6 in 400. Relative risks tend to act like a magnifying glass. As CBS also pointed out, they can make the numbers seem scarier than maybe they should.

Note that we have ignored—except by implication—what happens to the several hundred people in our sample who were unaffected before and are unaffected now, as does just about every news report of cancer risk.

Let's change this. Let's begin instead by focusing on those who do not or will not have this cancer, the people for whom the revolver hangs idly on the wall and is never fired, those for whom the fire goes out or is never lit. Now we notice the 395 out of 400 who don't ordinarily get pancreatic cancer. If all 400 eat the extra bacon every day for the rest of their lives, one more will have pancreatic cancer as before, but now we notice that 394 will still be fine.

Which looks rather less risky, don't you think? But we can make the numbers more striking by drawing attention to the fact that 399 out of 400 are unaffected by an almighty pig-out change in diet. There is, remember, only one extra case of this cancer, and the other 399 were either going to get the cancer anyway or were not and will be fine. Either way, 399 out of 400 cases are non-events as far as the extra bacon goes. Yet this is the same risk as the approximately 20 percent increase in risk that we found when we concentrated only on the events and ignored the non-events. Except that looked at this way, the change is not 20 percent but 0.25 percent.

The point is to ask if the way risk is usually talked about leaves the cup of danger pessimistically half-emptied by death, not half-full of life or survival. Only in this case the cup is 399/400ths full of life, or at worst 394/400ths—but the media tend to look only at how much it empties.

Perhaps that's what people want when they talk about danger. We want to talk about victims, not those to whom nothing happens. But this is a choice, even if we are unaware that we have one.

We could, as we say, talk about survival instead. And by concentrating on those who are unaffected, the danger in this case diminishes. It isn't even our focus anymore—living is—and the excellent prospect of survival is not much changed. We could describe the pancreatic-cancer effect of all that extra bacon every day for life as roughly a 99.75 percent chance of being unaffected.

In Figure 5, is the baseline risk of pancreatic cancer best measured by the 5 people in gray, who ordinarily get it (the events), or by the people in white, who are fine (the non-events)? Then, how do we represent the extra risk of pancreatic cancer if all 400 eat extra bacon every day, shown by the 1 person who is crossed out? Is that best expressed as an increased risk of 20 percent for the grays, or as a 0.25 percent decrease in the chance of being fine for the whites, or as a 99.75 percent chance of still being fine for the whites? It's all in the framing.

Although, following the work of psychologist Gerd Gigerenzer, there is a way to avoid percentages altogether, and talk about the pure number of people affected. Then, whether you want to talk about the increased risk or the change in the chance of being fine, the measure is the same: 1 person in 400.

The man who ate extra bacon every day and didn't get pancreatic cancer and was fine, just like almost everyone else, is a massively more faithful rendition of how life is. It is also "dull, dull, unbearably dull." Newspapers never write it that way. People don't tend to talk that way. But then, the whole concept of risk assumes a

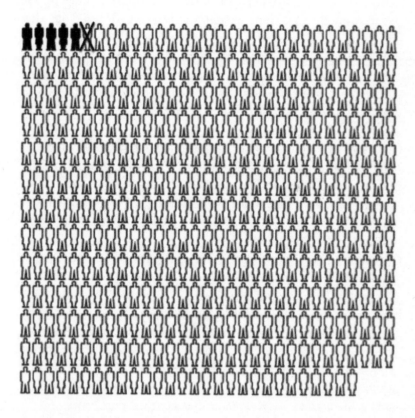

FIGURE 5. If 400 people all eat an extra 50 grams of bacon every day, with an associated 20 percent increased risk of pancreatic cancer, then one extra person will get this cancer during his or her lifetime.

"negative frame": risk is about the potential for bad, seldom the potential for life to go on, except by implication, an implication that must lurk in the background. Standing in the foreground is death. Is it any surprise which perspective holds our attention? So if cancer is shown hanging on the wall in Act I, then cancer frames our expectations. Thinking about what makes the risk go "up" tightens the focus further. We're mentally halfway to contracting it before we've even reached the last act.

Even MicroMorts do this. They, too, are about comparing the bad things that happen rather than the non-events. Perhaps they, too, need a complementary perspective, some sort of reframing. Maybe we need another new unit, an anti-MicroMort, to describe the daily chance of being fine. How about a MicroNot (1 MN)—a one-in-a-million chance that nothing fatal happens from a particular cause?

Then, if we take as an illustration the average daily dose of acute fatal risk—1 MicroMort—we can switch focus to the MicroNot to show that the average daily chance of an acute non-event—of being fine—is 999,999 MNs.

Then let's say that you do something, hypothetically, to double your average daily MicroMort risk from 1 MM to 2 MMs. Oh, no! It's up 100 percent! But reframe the risk with MicroNots to talk about non-events, and your MicroNot dose falls from 999,999 to 999,998, a fall in your chance of being fine of 0.0001 percent. Again, perspective, or framing, is all.

Another real news story illustrates a similar trick. Researchers found that a genetic variant—call it "X"—present in 10 percent of the population *protected* them against high blood pressure. Although it was published in a top scientific journal, the story received negligible press coverage until a knowing public relations director rewrote the press release to say that a genetic variant—call it "not having X"—had been discovered that *increased* the risk of high blood pressure in 90 percent of people.[2] This story was widely reported, whereas talking about those to whom nothing happened wasn't news.

Both good and bad endings are present in any risk of less than 100 percent; otherwise there is no risk, and the future is certain one way or the other for all. But the entire framing of risk is around the bad and changes to the bad, even when there is not much bad to be had, even when the good is overwhelmingly more likely and not much dented even by extreme changes in behavior.

Sometimes, but rarely, the bad is changed to the good. There was an advertising campaign on the London Underground that proudly declared: "99 per cent of young Londoners do not commit serious youth crime." Sounds wonderful. But, taken literally, that means 1 percent of young Londoners *do* commit serious crime, and there are around 1 million young Londoners, which implies 10,000 young thugs running around. Not so wonderful. The switch between events and non-events, between fires that burn and fires that go out, can have a powerful effect on people's decisions. DS has found that when statins are shown to reduce the risk of heart attack for people over-fifty by 30 percent, people are keen on statins. But they're not so keen when he also shows them that around 96 people in 100 will be unaffected if they take statins for ten years—they either would or would not have had a heart attack anyway—but might be vulnerable to side effects. The risk hasn't changed, only the way it is represented; but that changes people's minds.

Does this make them irrational? Some may think this is proof of human flakiness. Why aren't people consistent when the risk is the same?

We think that's unfair. A change in framing changes the context—that's what framing is. People unsurprisingly find it hard to weigh a risk against everything in life that's relevant. Shown a risk in a way that emphasizes danger, they react by saying it's dangerous. Shown it in another way that emphasizes being fine, and they say: "Oh, that changes things." Our experience is that, once they have seen it both ways, most people are more consistent. Some experiments that purport to prove irrationality through a change of framing seem more like a trick than a test. Do people who change their minds when the framing is changed change them back again if the framing changes back, and continue switching every time? Of course not.

In relation to stories, think of the numbers as follows: imagine 6 bad things out of 400 possibilities. This probability consists of a numerator and a denominator: 6 is the numerator and 400 the denominator. Narratives or stories often concentrate only on numerators. We tell stories about people who do things and people to whom stuff happens—the 6—not the people who do nothing and to whom nothing happens. This is almost the definition of what stories are. If you want to know about a story, you ask: "What happens?" And so we suffer from what is known as denominator neglect. We ignore the mass of people from whom our 6 happening examples are plucked.

Not all stories are so simple, it is true. Good fiction teases out and tests our expectations and sometimes frustrates them. It puts revolvers on the wall and leaves them there. It can be full of ambiguity. In 1925, Virginia Woolf distinguished her fiction from that of the past when she wrote: "If a writer were a free man and not a slave, if he could write what he chose, not what he must, if he could base his work upon his own feeling, and not upon convention, there would be no plot, no comedy, no tragedy, no love interest or catastrophe in the accepted style. . . . Life is not a series of gig lamps symmetrically arranged."[3]

The metaphor refers to the lamps on each side of a horse-drawn carriage, and so Woolf was implying, in part, that life proceeds in the dark, not with the foreshadowing of fiction. But this argument against orderly expectations has been used to exaggerate the modernist case. For it is also true that Hamlet's agonies about killing his uncle have exercised critics and audiences for four hundred years with their lack of symmetrical, gig-lamp clarity, as does Hamlet's every relationship. There are no signposting coughs in Hamlet, the gig lamps are not bright, and the play is brighter for it. Good fiction has played for centuries with causality and the limits of our knowledge.

But are the stories we tell in the news media or about our own lives as sophisticated? Life is at least as puzzling and messy as fiction. But do we, many of us, behave as if it were a bad novel, and try to impose more direction and coherence than many writers would dare? Risk is not, or should not be, a bad novel. In real life the gun on the wall often rusts. So maybe non-events should be bigger news. The *auteur* in Mitchell and Webb's sketch is that rare and uncelebrated thing: an artist of the denominator. Give him an Oscar.

All this is part of a general problem with risk perception known as availability bias, although, as we say, "bias" is a mean word for what is often just a result of the framing. Availability bias refers to the fact that some things come to mind or attention more readily than others: the gun-on-the-wall effect. And sure enough, it's easier to think about events than non-events. Whatever it might mean to try to think about an event that doesn't happen, it is not an easy or intuitive thing to do.

Researchers in the 1970s, among them the psychologists Daniel Kahneman, Amos Tversky, and Paul Slovic, ran dozens of human experiments to discover what influenced people's estimation of risk.[4] They noticed that after a natural disaster people took out more insurance, then, with time, took out less. This was not because the risk rises immediately after a disaster and then falls, obviously, but because the risk is more salient immediately after a disaster.

Slovic also found that tornadoes were seen as more frequent killers than asthma, although the latter caused twenty times more deaths. As we will also see in the chapter on crime, vivid events are recalled not merely more vividly but in the belief that there are more of them. Put crudely, we worry more that something might get us not because it's more likely to get us but because it would make better television.

This is a bias that can strike en masse, especially if we frame it right. The media play a part, and Kahneman uses the term "availability cascade" to describe the surge of interest, attention to new cases, and panic that come with a scary but rare event. In contrast, problems that are common are not surprising and are less likely to qualify as big "news." Another smoking death? And? So things that are genuinely likely to get you are not reported nearly so often as others that are rare. The unusual is, by the nature of news, disproportionately reported, so we think it more common.

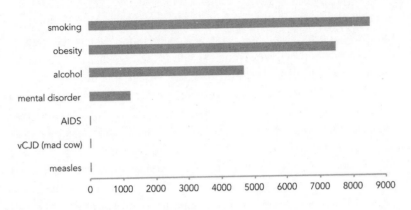

FIGURE 6. Number of deaths from each cause in one year for each story about this risk that appeared in BBC news.

Although this is a reporting bias, the media have no trouble justifying it on the grounds that people want to know about what's unusual and new, not what's old. There is no way they could report risk proportionately and still be in business. It would mean thousands of times more articles on smoking than on death from measles. But it is a bias, and although it is only speculation that this reporting bias affects people's estimation of the size of different risks, it seems like reasonable speculation. The extent of the bias has even been quantified (see Figure 6).[5] Mad cow disease, measles, and AIDS received massive coverage in the United Kingdom relative to the number of people who died, whereas smoking, obesity, and alcohol received little. The precise numbers of deaths attributable to all of these is debatable, and the fashion for certain stories has changed since the research was carried out more than a decade ago, but the thrust of the argument seems fair.

So there are all kinds of reasons why risk depends on what we pay attention to. This is partly because our attention is scattered, but it's also because it's not clear which version of the numbers is the one to watch: the numbers about what happens, or the numbers about what doesn't.

ACCIDENTS

They were eleven, Norm and Kelvin, the time they went to swim in the reservoir. It was warm as they pedaled along the lanes, and when they arrived the water looked cool. They threw down their bikes by the reeds, took off their shoes, stripped to their briefs—Norm's blue, Kelvin's white—and stood on the grass bank not far past the two fishermen. The dare should have been a breeze, to be honest, but Norm hadn't figured on the wind-up.

"No, really, they do!" said Kelvin. "They're that long, with these evil teeth, and they creep up underneath and go [snapping his jaws together] gnah!"

"Yeah, but not your bollocks," said Norm.

Kelvin made rabbity, nibbling movements with his top teeth against his bottom lip, leaving brief pale slots in the pink skin.

"No way!" said Norm, looking away.

"Big. And pointed."

Kelvin paused. Norm felt the breeze.

"Chew your knob off."

"Shut up."

"Here, pikey-pikey, Norm knob!"

"*Shut up!*"

For a moment there was only the water beating at the reservoir wall and the wind in the reeds.

"What's up?" said Kelvin. "Cold?"

"Nah. You?"

"Nah."

Norm half-folded his arms, then let them drop.

"So . . . scared, then? Or what?"

"Me? Scared?"

"Go on then."

The water chopped, and the waves slapped.

"Dive, yeah?"

"Yeah, 'course, dive."

"You do it then."

Danger isn't a choice at the age of eleven; it's a game with rules. Kelvin stood on the edge looking at Norm over cupped hands. Then his arms flew back, his body was a breath and a twitch, his white legs flexed and leapt.

Kelvin was in, smashed by cold, then above the water, hitching up his pants, head high, blowing at choppy peaks across the reservoir. Norm twitched, too. But his body refused. His feet held to the ground.

Norm picked up his clothes and held them to his chest, eyes on the dark head in the waves, his thoughts on the precious last inch of cold flesh in his underpants and the imagined snap and rip of a pike.

On the outside, Kelvin looked hard and fearless. On the bank, Norm was transfixed by a toothy, jelly-eyed monster like something he'd seen in a fish shop, staring at him close-up. As Kelvin pulled through the murk toward a soaked wooden platform on the far side, Norm cupped a hand between his legs.

Kelvin's swim to the platform would have been cold. Once there, climbing out seemed stupid as soon as he'd done it, except that Kelvin did it and Norm didn't. Then what? Swim back. Colder still.

As Kelvin swam his last, slower strokes, Norm looked down, one hand clutching his clothes, the other still down his underpants. Out of the corner of his eye he noticed the fishermen again. A shout, "Oi!" turned into a clatter of boots. Running.

Norm looked up: flapping boots, chopping water, running and shouting. He turned back to see Kelvin climb out, wherever that was. He looked left then right, up and out over the water for the pale body and the dark hair.

"Kelvin?"

He twisted round. Looked back again for the place he'd missed. He saw water, green with flecks of gray and low, chopping waves, the platform, the bank, Kelvin's clothes.

"Jesus," he mumbled and stared again.

"Kelvin!"

But there was no hair, no skin. Instead, his mind filled with the jelly eye and the sharp teeth, the eye, bite, and blood. Fishermen plunged past into the water and the reeds hissed in the breeze.

He was white when they found him. But there was no blood. Still had bollocks, too. The water did it, or maybe the cold. He wasn't far out, or very deep. Three more strokes, they said. He should have shouted, but the fishermen said he just went under. It wasn't Norm's fault, adults said.

Later, Norm felt vaguely disappointed that the big thing hadn't happened and his friend didn't die. Not even brain damage, they said. Then Norm felt vaguely guilty for feeling vaguely disappointed. At school on Monday it was cool to be Kelvin. Norm would have liked to be cool, too, but he was still thinking about teeth and jelly eyes.

· ·

THERE ARE TWO WAYS to judge the risk of being bitten by a pike. One is, "Don't be daft, what pike?" Even if there are pike, they probably have zero appetite for Norm's willy. On the other hand, we could say the risk is stratospheric—because, come on, this is his willy he's thinking of. After that, what more needs to be said? The risk is beyond imagining.

The difference in these two calculations of the same risk is that between probability and consequence. Probability describes the chance of being bitten; consequence is the answer to the "what if?" question. Is it likely he would have been bitten? Who knows? Probably not. But what if . . . ? Same risk, different perspective, sea change in emotion and anxiety. Norm's struggle on the bank is between long odds and imagined agony. No prizes for guessing which grabs his attention. Just follow his left hand.

Fatal accidents are often like that: rare but terrible to think of. It is rare for children of Norm's age to drown: about 1 or 2 in every million children in the United Kingdom between the ages of 5 and 14 die through "accidental drowning and submersion" each year, and about 3 or 4 in a million for those younger than 5. The figures are higher in the United States, presumably because there are more private swimming pools to fall into: around 6 per million for children aged 5 to 14, and for those under 5 a frightening 475 drownings, representing around 25 MMs per year per child.[1] There are no official figures for death by drowning by pike.

Even on the roads—where the threat of an accident is perhaps most apparent to parents (like us) who have clutched their children's hands—it is increasingly rare for children to die. But all this is little consolation if you think of consequences instead of probabilities. One in a million, as Prudence's mother said in Chapter 2, is useless to the one.

Take the true story of Mark McCullough, the father of a seven-year-old girl who allowed her to walk the 20 yards from their home to the bus stop on the way to school each day, unaccompanied. Naturally, reported the British *Daily Telegraph* in September 2010, he was threatened by a council with "child protection issues."[2]

The walk meant crossing a "busy road" (said the council) or "quiet country lane" (said Mr. McCullough). The council was probably "only doing its job." Mr. McCullough was "very angry." "It's an absolute joke," he said: "I am not going to wrap my children up in cotton wool." A fact confirmed one chilly morning when his daughter was spotted "without a jumper," said the council's shocked head of transport services.

As with drowning, the figures for the kind of accident the council fears are low. In 2008 in England and Wales there were 1,471,100 girls between the ages of 5 and 9. The UK Office for National Statistics (ONS) says that 137 of them died from all causes. Seven died in traffic accidents. One was a pedestrian.*

---

* In 2010 there were no pedestrian deaths in this category.

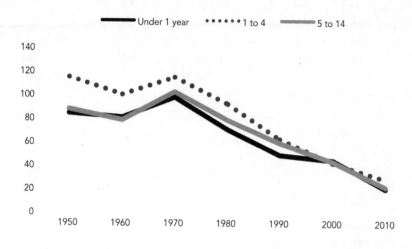

FIGURE 7. Children killed on the roads in the United States, 1950–2010, in MicroMorts per year.

This is an average risk of less than 1 MicroMort per year: that is, 1/365th of the average daily risk of acute death for a UK adult. As we show below, more children die each year in the United Kingdom by being strangled by cords on window blinds. US roads appear a bit more dangerous: there were 25 pedestrian deaths out of 10 million girls this age in 2010, representing around 2.5 MMs per child per year.[3] If we look at the risks of all US child deaths on the roads, whether the child was walking, on a bicycle, or riding in a car, the graph shows an amazing decline in 60 years—as cars get safer, children are more protected, and trauma care gets better.

Put aside arguments about parental responsibility and freedom. Concentrate on the risk—and answer an awkward question: Which are most compelling, the basement-level mortality statistics, or the heightened fears of Lincolnshire County Council?

If you back the council, it may be because you feel the pull of what's called the "asymmetry of regret," better known as "How would you feel if? . . ." How would you feel if you took your child to and from the bus stop every day and your worst regret was that you wasted a few minutes to prevent an accident that was probably never going to happen anyway? Not entirely happy, maybe, but in the scheme of things the time and effort are no big loss. Next, how would you feel if you did not give up those few minutes, then one sunny morning over toast and marmalade you heard the squeal of brakes?

Choosing what to do in life by trying to minimize what you might most regret is known in decision theory, naturally enough, as mini-max regret.[4] Most decisions are here and now. Imagined future regret is a tortured complication. It's more than the difference between short-and long-term self-interest. It twists the knife with potential guilt and blame, imagines this as if experienced with hindsight, then makes that anticipated-retrospective feeling our main motive, if you follow. It is a roundabout way of deciding what to do that people can calculate in a blink. But at its most tyrannical, mini-max regret makes us slaves of nightmares. A bad ending foreseen or imagined comes to justify the most drastic avoidance. Does it also save lives? Thomas Hardy said that fear is the mother of foresight. In the TV series *The Simpsons*, "just think of the children" is a joke that could be about mini-max regret. For some, it is a compulsion.

For people who feel this way, it is the potential "what if?"—maybe the maximum imaginable "what if?"—that they will do just about anything to minimize, and that weighs far more than any odds. So it's no surprise that probabilities sometimes cut no ice, no matter how low the chances are of the feared result, especially if the emotional stakes include children.

Still, let's remind ourselves—again—that the probabilities are lower than ever. As with mortality overall and also with murder, children are far less likely than any other age group to suffer accidental death.

Old age, not childhood, is by far the most dangerous time for avoidable accidental death. The accident rate for US males who have gotten to 85 is so high that it is cut off at the top of the graph shown in Figure 8.[5] We couldn't fit it on the graph without making the risk to most other age groups vanish into the axis. For women, the clear trend is that death from accidental injury becomes more likely with age. For men, the trend is not quite so smooth, bucked in early adulthood and middle age by a taste for thrills that are bound to cause a few casualties.

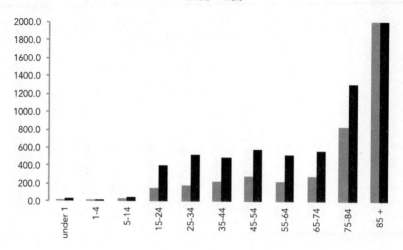

FIGURE 8. MicroMorts per year for accidental injury (including transportation) in the United States, by age, 2010.

If we highlight in these overall figures just the accidental deaths related to transportation, it's a slightly different story for young people. The old are still among the most vulnerable, but those aged 15 to 24 are now up among the highest, for fairly obvious reasons.

But, as before, children under 15 years old are far less likely to die in a traffic accident—which includes being hit by cars—than any other age group.

Many risks to children have fallen sharply over the past 50 years. You've probably met the carefree-youth bore with his "I-used-to-roam-around-all-day-and-we-got-up-to-all-sorts-of trouble-dodging-traffic/bullies/the-local-flasher-and-it-never-did-me-any-harm." Rose-tinted memories of childhood are a constant. Two out of two of the authors of this book might even have been caught expressing them. But such people are often wrong. Many changes are clear-cut, and road accidents are a good example.

Watching a British postwar Central Office of Information film, with men in hats and women in coats with big shoulders, reminds us how much roads have changed. In 1951 there were fewer than 4

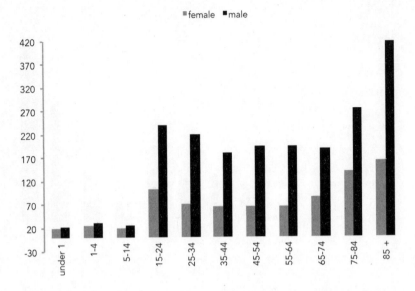

■female ■male

FIGURE 9. MicroMorts per year for transportation-related accidents in the United States (equal to crude rates of death per 1 million population), by age, 2010. Note the different scale from Figure 8.

million registered vehicles on the roads in Britain. They meandered the highways free of restrictions such as road-markings, traffic calming, certificates for roadworthiness, or low-impact bumpers. Children played in the streets and walked to school. The result was that 907 children under 15 were killed on the roads in 1951, including 707 pedestrians and 130 cyclists. Even this was less than the 1,400 a year killed before the war.[6]

The carnage had dropped to 533 child deaths in 1995, to 124 in 2008, to 81 in 2009, and in 2010, to 55—each a tragedy for the family, but still a staggering 90 percent fall over 60 years. Over the same period, the number of registered vehicles went up more than eightfold, to 34 million. If we assume a steady decline from around 1,000 deaths in 1951 to 50 now, this means an average of 450 children's lives have been saved in each of those 60 years: that is 27,000 people alive today in Britain who would have been killed on the roads as children if accidents continued as they had in 1951. As shown earlier in Figure 7, the fall in the United States has not been quite as dramatic, but is still impressive: 98 of every 1 million children from the ages of 1 to 14 were killed on the roads in 1950; this figure had dropped to 23 by 2010, a reduction of 77 percent in risk.[7]

And it's not just that traffic and cars are safer. There have been huge improvements in emergency medical treatment that mean children in accidents, who in the past would have died, now run around to get knocked down another day. This is shown by the fact that injuries have dropped, but not as fast as fatalities: in Britain in 1951, 5,743 children under age 15 were seriously injured on the roads. This number went down to 2,502 by 2010, a fall of 56 percent, though there is also much argument about the accuracy of the injury figures. One figure has not changed: around two out of every three child casualties are male.

But roads are not the only hazards for children. The ONS reports that in total in 2010, out of 9.5 million children under the age of 14 in England and Wales, 172 died of accidents. This included 21 pedestrians, 12 cyclists, 17 car passengers, 22 by drowning, 27 by accidental strangulation, and 10 in fires.[8]

This is an average for each child of 18 MicroMorts per year from accidents, which seems rather safe compared to the situation faced by kids in the United States, where, in 2010, 4,147 children were killed in accidents out of 66 million aged 14 or under, including 1,400 on the roads, 726 by drowning, and 62 in "firearm accidents"—that works out as 63 MMs a year, over three times the UK risk.[9]

It is perhaps remarkable that about the same number of children in England drowned as were knocked down as pedestrians. There were also more accidental strangulations than drownings or pedestrian deaths: looped window-blind cords are now recognized as a particular danger for toddlers around 2 years old, and in 2010, IKEA withdrew more than 3 million blinds for this reason.[10] But if you are a parent, of which risk are you most afraid?

This huge reduction in accidental deaths and injuries must be good. Who wants more danger for children? But it is not always easy to know why children are not killed and injured on roads. Maybe a big reason is just that they are not outside walking to school or playing. So improvements may come at a price. We can report the numbers; we can't altogether explain them or their consequences. In 1971, about 80 percent of 7- and 8-year-olds in the United Kingdom traveled to school without an adult. By 2006, the figure had dropped to 12 percent of 7- to 10-year-olds, with nearly half of all primary-school children arriving by car. In 1971, the average age at which children were allowed to visit friends or shops on their own was 7. In 1990 this freedom was on average first granted to 10-year-olds.

These statistics reflect a growing risk-aversion on behalf of children, detailed by Tim Gill in his 2007 book *No Fear: Growing Up in a Risk Averse Society*.[11] Gill suggests a number of possible reasons for this trend: more cars, less open space, computer games, working parents, and the greater publicity given to accidents and tragedy. Clichés such as the "nanny state" and "compensation culture" come easily to mind, but the issue is complicated.

Take playground safety. Some postwar playground equipment now looks as if it were designed to injure children: even children of the 1950s knew to watch themselves on the witch's hat and swingboat. In the 1970s, even before ideas of the nanny state and compensation culture took hold, the populist and popular BBC TV show *That's Life* ran a campaign for safer playgrounds, demanding new equipment and resurfacing. As Gill points out, these expensive changes led to a reduction in playgrounds, which initially had the unintended consequence of encouraging children to play in the streets, and little benefit in terms of reduced injuries. This may have been through the phenomenon of risk compensation: DS saw this in his child's primary school when some large, ancient, much-loved piece of rotting wooden playground equipment was replaced by a bright, new, and rather dull climbing frame: in the first week, a child tried to extract more excitement by balancing on the top bars, fell off, and broke her arm.

Other countries have not been so obsessed with playground safety, and standards are now being relaxed rather than strengthened.

The problem is that it is easy to quantify the danger of allowing, or even encouraging, children to have adventurous play, and not so straightforward to quantify the benefit, which is hard to prove.

Gill argues that danger can be good for kids; they have an appetite for risk and will seek it out one way or another, so it is better to teach them how to manage risky situations in the future, such as from water or traffic, without getting into trouble. This benefits healthy development; it is, to use another cliché, character-building; it encourages a sense of adventure, entrepreneurism, self-reliance, resilience, and all of those other fine characteristics on which Western Civilization is supposedly built.

You often hear stories of people putting a stop to any activity that seems remotely fun by citing "elf and safety" to avoid the charge of failing to look after children. Even the UK Health and Safety Executive thinks they go too far.

If that's not what you expect of the HSE, perhaps you've heard too much from its critics. But its public statements could have been written by Tim Gill. One HSE document includes a statement that sounds like Mr. McCullough: "No child will learn about risk if they are wrapped in cotton wool." The HSE goes on to say that it does not expect that all risks must be eliminated or continually reduced, or that every aspect of play provision "must be set out in copious paperwork as part of a misguided security blanket."[12]

We look more closely at health and safety risks in Chapter 18. The point here is that some low numbers for child accidents might indicate overprotection, which children pay for when they're older but no wiser, an argument the HSE appears to accept.

The chair of the HSE, Judith Hackitt, has said: "Don't use health and safety law as a convenient scapegoat or we will challenge you. The creeping culture of risk-aversion and fear of litigation . . . puts at risk our children's education and preparation for adult life."

The Countryside Alliance, a British organization promoting sports such as hunting and fishing, found that in 138 local authorities in England, over a period of 10 years, there were 364 legal claims following school trips, of which 156 were successful, resulting in an average total annual payout of £293 (US$500) per authority between 1998 and 2008. Compensation culture gone mad? In the 5-year period between 2006 and 2010, the HSE brought only two prosecutions involving school trips.[13]

Occasionally, the bad side of over-anxiety becomes clear. Obsessive use of sunscreen on children, or lack of outdoor activity in the sun, can lead to vitamin D deficiency, with the result that rickets, once thought to have been almost eliminated, is becoming an established disease of Middle England. Kellogg's even proposed to fortify its children's breakfast cereal with vitamin D, citing research that showed a 140 percent increase between 2001 and 2009 in the number of British children under 10 years old admitted to hospitals with rickets. A recent study from the Mayo Clinic in Minnesota concluded that "nutritional rickets remains rare, but its incidence has dramatically increased since 2000."[14]

Tim Gill calls for a change from a philosophy of protection, in which every accident is seen as someone's failure, to a philosophy of resilience, meaning the ability to thrive in a world in which bad things happen.

As he says, "childhood includes some simple ingredients: frequent unregulated self-directed contact with people and places beyond the immediate spheres of family and school, and the chance to learn from their mistakes."

Although children in the United Kingdom and in the United States are more protected from accidents than ever before, the same cannot be said of all children everywhere. Road traffic accidents are the second most common cause of death of children between the ages of 5 and 14 worldwide, and the most common cause for 15- to 29-year-olds. An estimated 240,000 children a year are killed on the roads in developing countries. When we venture overseas, even those who encourage kids to walk to school at home will shudder at the sight of small children in spotless uniforms, walking to school by the side of roads with thundering traffic and no protection.

VACCINATION

She saw Prudence turn again in the dark. Vile, black water streaked with acid poured across the girl's path. From nowhere, gray beasts with clanking teeth began to prowl while stinging insects steamed from the undergrowth, hungry. Nettles groped for her skin. Prudence hid her eyes and backed away—and stumbled into thorns that ripped her arms with poison as she fell to the forest floor where creepy-crawlies smelled blood and skittered over her legs to pierce her with pincers. And as Prudence writhed and screamed, pricked, bitten, and burned, her mother watched, lifted her foot, and pressed a stiletto into her daughter's back to hold her down.

Everything a mother does is wrong, she thought, when she woke, disturbed, and later read that the dangers of life are infinite and among them is safety, according to Goethe, apparently. "Thanks," she said.

"Just a little scratch," she said to Prudence after breakfast. "Everyone has it. You can have a Band-Aid."

Because if safe could be dangerous, and the dangerous things were only allayed with more danger, which was really safe, and how risky was safety? Or whatever. Bloody Goethe. Bloody advice.

"It's to protect you," she said, as Prudence sat in her Tripp Trapp chair, coloring with little arms.

On the kitchen table lay printed pages from the Student Vaccine Liberation Army, "armed with the most powerful knowledge" to encourage critical thinking about "the risks and dangers of vaccines," which explained how "in England in the earlier centuries hatters used to use mercury to stiffen the felt of hats. The mercury was absorbed into their fingertips and they eventually went "mad" . . . MAD AS A HATTER. We are injecting mercury directly into our children's arms causing an epidemic of apparent madness . . . ADHD, ADD, Autism, OCD, Bipolar . . . are vaccine injuries mis-labelled as mental illness."[1]

A leaflet from the doctor said: "All medicines (including vaccines) are thoroughly tested to assess how safe and effective they are. After they have been licensed, the safety of vaccines continues to be monitored. Any rare side effects that are discovered can then be assessed further. All medicines can cause side effects, but vaccines are among the very safest."[2]

Bad if you do, bad if you don't, bad to Pru, bad to others. Preying on her mind were vaccination horror stories about children who died or suffered because they were not protected from common illnesses. "If only we had given him the shot," they said. If they had only vaccinated their own child, their friend's baby would not have been exposed to the virus before vaccination and would still be alive.

On her laptop was another mother's story: after shots for hepatitis her baby screamed incessantly, hardly slept, didn't feed, woke one morning vomiting, was airlifted to the hospital, survived with "seizure disorder, cortical blindness, severe reflux and high risk for aspiration pneumonia . . . severe developmental delay . . . a mixture of hypotonia . . . [and] some spasticity," needing twenty-four-hour care by two people.[3]

On the news last night was a report of a warning sent to schools about measles abroad, the risk of traveling if not immunized and the potential complications: eye and ear infections, pneumonia, seizures, and, more rarely, encephalitis—swelling of the brain, brain damage, and even death. On top of everything else on the table was an appointment card.[4]

In the end she went, telling herself to calm down but helpless with guilt, and for three whole days her heart beat at her conscience while she watched, and waited.

. . . .

WHAT IF PRUDENCE had the vaccination and then soon afterward began to show signs of autism? Is the story then that her mother's decision to vaccinate was catastrophic? The truth is, we don't know exactly what causes autism—although there are plenty of theories. But when one thing follows another, the easiest story to tell is that one caused the other. So whatever she does, Prudence's mother feels she will be in the dock for whatever happens, as storytelling joins the dots in hindsight with any emotional glue, fear, politics, prejudice, or presumption that come to hand to tell us that she got it wrong. Pity the mother, who can't win against risk when stories come to be told after the event.

Pity the small boy, too. When DS was a child, there were no vaccinations against measles, mumps, and chickenpox, so when anyone local had them he was marched round to be infected. The rule of thumb was that, if you were going to get it, it was best to get it over with now. Vaccination is a set of uncomfortable trade-offs, because DS now knows that measles exposed him to around 200 MicroMorts.* But there wasn't much choice back then, and no doubt it was character-forming.

Measles was common in the 1950s, the effects well known, complications not unusual. It can cause blindness and even death. Observe that in your neighbors' children and you cry out for protection. Because it was common, people saw it around, a clear and present danger they wanted to get rid of. Now that we don't see it so much, some cry instead for protection from vaccines. Visibility plays a big part in risk.

---

* Between 1958 and 1960, when DS was around 6 years old, 957,000 cases of measles were reported in the United Kingdom, of whom 178 died.

Given a choice between harm you can see and harm you can't, what would you do? Some avoid the one they can see—it seems more threatening than a danger that's not here and now. Some take bigger fright at invisible risks, such as radiation, which seem more sinister. Others try for a coolheaded calculation of the pros and cons. And some of us simply do what the doctor tells us.

The simple point here is that "risk" is more typically about "risks," plural, and these risks often point in different directions: some immediate, some distant, some visible, some latent. How are you supposed to decide which risk to take when you are often judged not by the fact that you did your best in all good faith in a state of uncertainty but by how it turned out? Pru's mother sits at the table beset by tragic endings.

If she types "vaccine safety" into her browser, she will see mostly official sites full of reassurance. If she types "vaccine risk," she will find claims that children are harmed, that the science is bunk, and that the scientists are not to be trusted.

Vaccination brings out many of the fear factors that arouse strong emotions (see also Chapter 19, on radiation). Children are not even sick before we inject them. Actively sticking a needle in your child's arm is a sin of commission that hurts in ways that doing nothing doesn't—at least, not right away.

Vaccination is also imposed, either through pressure or legal compulsion: if your child is to attend a kindergarten in Florida—or most other states—for example, they must have been vaccinated against diphtheria, tetanus, and pertussis, or whooping cough (DTaP); hepatitis B; measles, mumps, and rubella, or German measles (MMR); polio; and varicella (chickenpox).[5] There can be side effects. Finally, multinational corporations make a lot of money out of this mass-medicalization.

All of which can provoke fierce opposition. And so claims that vaccination may cause dread outcomes such as autism find a ready audience, particularly in the United States.

At least children don't need to be vaccinated against smallpox anymore, since the disease was eradicated after the last natural case appeared in Somalia in 1977 (although a UK laboratory worker died in 1978). Smallpox killed countless numbers throughout history, with 2 million deaths a year right up to the 1950s. It helped Europeans conquer the Americas by almost wiping out the native populations. But it had long been observed that survivors did not contract the disease again, and the practice of *inoculation* developed, in which extracts from scabs were scratched onto the skin to give a deliberate, but hopefully mild, infection.

Then, in 1796, Edward Jenner used another, much milder disease, in this case cowpox, after it was noticed that milkmaids tended not to have smallpox. Hence the term *vaccination*, from the Latin *vacca* (cow). . Being scraped with pus taken from someone who had been exposed to an infected cow's udder isn't an obviously attractive idea, so naturally it was tried on a small boy—James Phipps, aged eight, the son of Jenner's gardener. Whether he gave informed consent is not recorded.

Moving vaccine around in the early days was tricky without refrigeration. The answer, as ever: small boys. To ferry cowpox to the Spanish Americas in 1803—a voyage so long that one infected person would recover—eleven pairs of orphans were press-ganged into service and the first pair infected before they set sail. In time, they passed it on to the next pair, and so on for the duration of the voyage until good, fresh cowpox pus arrived in the New World. It took a while to catch on, but immunization (which covers both inoculation and vaccination) went on to save millions of lives.

The history of measles offers a clue about the risks without immunization. In England and Wales in 1940 there were 409,000 cases, of whom 857 died, a "case fatality rate" of 0.2 percent, or 2,000 MicroMorts, the same as that quoted by the US Centers for Disease Control and Prevention (CDC) in the United States. Vaccination started in the 1960s, and by 1990 the number of cases had dropped to 13,300, with 1 fatality. Since 1992 there have been no childhood deaths from measles, but only adult deaths resulting from rare late effects of infection acquired earlier in life.[6]

Vaccination can also be used to reduce long-term harm: it's been estimated that one cancer in every six is caused by infection, but the long period between infection and illness makes the association hard to spot. Nevertheless, the human papillomavirus (HPV) vaccine is now offered to 12-year-old girls and boys in the United States to help prevent the infection that can lead to cervical cancer.[7]

So it seems a good thing to be vaccinated. Rather like stopping smoking, it is also good for people around you. This is because of herd immunity, which occurs when sufficient numbers of people are immune for an infection not to turn into an epidemic. What "sufficient" means depends, in a fairly simple way, on how infectious the disease is. For example, a single person with smallpox would infect, in a community of susceptible people such as the Incas, on average around 5 others.[*] If they, in turn, infect 5 each, and so on, then it will only take 6 more steps to infect an entire community of 50,000 people.

Given the infection rate for measles, we need to vaccinate 92 percent of the population to prevent the spread of an epidemic.[†] This

---

[*] The average number of people infected is called the *basic reproduction number* and given the symbol R0 (R-nought)—for smallpox it is around 5, for measles around 12.

[†] Suppose the community had been vaccinated, to the extent that more than 4 out of 5 people (80 percent) were immune. In this case, someone with smallpox would on average only infect less than 1 new person, and so the epidemic would die out. So we get the nice general result that an epidemic will die out provided that of every R0 people, at least (R0−1) are immune, so the proportion that need to be immune is (R0−1)/R0. For example, to prevent a measles epidemic, we calculate $(12−1)/12 = 11/12$, showing that 92 percent of the population needs to be immune $(11/12 = 0.916)$, whereas the swine flu virus in 2009 was a lazy thing whose R0 was only around 1.3, and so only 23 percent had to be immune to stop the epidemic $(0.3/1.3 = 0.23)$. Since vaccines are not 100 percent effective, in practice the aim is to vaccinate a greater proportion than this, at least 95 percent in the case of measles. Of course, it's possible to get a "free ride" by not having the vaccination yourself, and relying on everyone else to stop an epidemic.

The English vaccination rates for measles in 2011 stood at 89 percent, up from 80 percent in 2003 but still not back to the 92 percent of 1995, let alone the 95 percent recommended by the World Health Organization (WHO). National Health Service (NHS), United Kingdom, Health and Social Care Information Centre, "NHS Immunisation Statistics, England 2009–10," 2012, available at www.ic.nhs.uk/Article/1685. After an outbreak of measles in Liverpool in February 2012, the Health Protection Agency revealed that 7,000 Merseyside children under the age of five had not had their full measles vaccine. Measles is the first M in the MMR vaccination, and coverage went down after the highly publicized claim

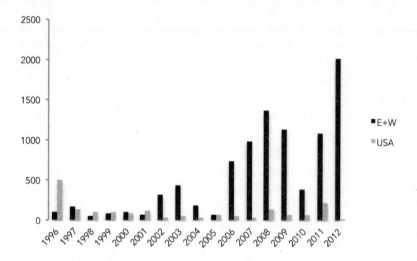

FIGURE 10. Confirmed cases of measles in England and Wales and the United States, 1996–2012. (US population is about six times larger, but cases are far fewer.)

has been achieved in the United States, with states having an average coverage with two doses of the MMR vaccine of 95 percent. Even so, in 2011 there were 222 measles cases reported in 17 outbreaks, mostly imported cases in unvaccinated people. In England, the rate dropped to 80 percent in the mid-2000s, although by 2011 it had recovered to 89 percent. Figure 10 contrasts what has happened to cases of measles in the United States and England and Wales, graphically warning of what can happen when vaccination rates drop.[8]

The fact that any one of us can have most of the benefit of vaccination without actually being vaccinated is known as a "free-rider" problem. You can rely on other people to keep the risk low—as long as they don't stop too.

So, is it risky not to be vaccinated? It is, and it isn't. The risk could be nil, and it could be huge. That is because it depends on

---

in 1998 that MMR was associated with autism. This claim has now been discredited, although it continues to have strong supporters in the United States—just try searching online for "vaccine autism."

what other people do as well as on what you do. If they all carry on vaccinating and you stop, you'll probably be fine. If they stop, too, you could be in trouble. The risk is dynamic and contingent. We are both subject to the risk—from others—and we are the risk, because we could become infectious.

So exactly the same behavior on our part in this case can mean extreme differences of risk, depending on what other people do. This makes reliable numbers about the disease risks that you face as an individual impossible to calculate. There's a big risk down the road for all of us if there is a failure of herd immunity, but what that means for you, now, is impossible to say. You could be fine. But what if you contribute to a loss of herd immunity? You could be dead. So, what's the risk?

But there is no denying that vaccines can have side effects. For example, the UK Medicines and Healthcare Products Regulatory Agency (MHRA) published the adverse-event reports it received following at least 4 million doses of the HPV Cervarix vaccine. There were 4,445 reports listing 9,673 reactions, although these reports are voluntary, and so the rough rate of 1 report per 1,000 injections is a major underestimate. The US CDC warns of mild to moderate reactions in 1 in 2 cases. Most of the MHRA reports are minor consequences of the injection, such as pain and rash, and over 2,000 were considered "psychogenic," caused by the injection process rather than the vaccine itself, including dizziness, blurred vision, and cold sweats.[9]

The problem comes with rare but severe events that occur later, and it is debatable whether these were due to the vaccine or would have occurred anyway. The MHRA lists over 1,000 reported reactions that are not recognized as associated with the HPV vaccination, including four cases of chronic fatigue syndrome. Given the number of 12- to 13-year-old girls in the program, the MHRA estimated it would expect to see 100 new cases of chronic fatigue syndrome over this period anyway, regardless of vaccination, so it is remarkable how few cases have been reported. But those families may well be convinced that the vaccine caused their child's condition. It is the "available" thing to blame, and didn't one thing lead to another?

The real problem is that with any mass intervention there will always be bad occurrences around the time of the shot—essentially, coincidences. For example, in 2009 a story in the British *Daily Mail* declared that a schoolgirl had died after receiving an HPV vaccination. The story quoted the school's head teacher as saying, "During the [immunization] session an unfortunate incident occurred and one of the girls suffered a rare, but extreme reaction to the vaccine." Three days later it was revealed that the girl had cancer and the death was coincidental; however, this was not headline news, and the tragic event has been used repeatedly on websites as proof of the dangers of the vaccine.[10]

But sometimes the reports are real. There was a classic example in 1976, when a new strain of swine flu was identified in Fort Dix, New Jersey. Fearful of a repeat of the 1918 flu epidemic that killed millions around the world, the authorities ordered mass vaccination, and 45 million people were immunized.

Within a year the program was abandoned, for two reasons. First, around 50 cases of Guillain-Barré syndrome—a gradual paralysis that is now thought to have been Franklin D. Roosevelt's condition—were reported, and 500 were eventually identified among those vaccinated. This suggested that among every million who had the vaccination, 10 more people than usual would get Guillain-Barré. In all, 25 died.[11]

The second reason the program was stopped was that the epidemic never got out of Fort Dix: nobody else had the flu, and there seemed to be no benefit to balance the possible harm. The director of the CDC was later sacked but still believes the vaccination program was right.[12]

Not all flu vaccines have the same risks. Following the UK swine flu outbreak in 2009, 9 cases of Guillain-Barré syndrome were diagnosed within six weeks of vaccination, but it was concluded this would be expected by chance alone.[13] But Finland and Sweden have reported increased rates of narcolepsy—sudden paralysis and sleepiness—in children after the swine flu vaccination, and this is still being investigated.

As the MMR saga has shown, it's hard to disprove an association. Thimerosal is a preservative used in some vaccines and contains mercury. It has long been accused of harming children. The CDC says there is "no convincing evidence of harm," but it was agreed in 1999 that it should be "reduced or eliminated in vaccines as a precautionary measure."[14]

The official line that the overall benefits of vaccination outweigh any risks ignores the way in which imposed and highly visible harms, however rare, are seen very differently from potential benefits sometime in the future, which can never be confirmed and seem "virtual" in societies where the risks of infectious diseases are so low.

It is a different matter in less-developed societies: the World Health Organization reports that there are still 140,000 deaths from measles each year, one every 4 minutes.[15] And, as we have seen in England, these are preventable. Vaccination has already made huge inroads: there used to be 2.6 million deaths a year. The eradication of measles is thought to be feasible, just like smallpox, especially now that the vaccine is stored in the fridge, not in a small boy.

COINCIDENCE

Fog everywhere. Fog lying in city streets like sleeping vipers. Fog, ghostly and concealing. The fog of novels, thick, bleak, and secretive. The fog of spies, mysteries, and thieves.

Norm, eighteen years old, walking. Hesitant in the gloom. Lost in thought of his father, who set sail two years ago to the day in his adored sloop Bill, for a sally off the south coast of England to Cowes on the Isle of Wight—as was often his habit—and in the most perfect of English zephyrs disappeared, boat and all, never to be seen again. How Norm yearned to hear once more the old man's rambling seafarer yarns. Even the fog seemed to whisper memories of the sea.

As he stole through the damp, gray air, Norm's left foot chanced to strike upon something soft. He was of a mind to walk on but fancied he heard a moan, and the humor came upon him to investigate, whereupon he discerned through the fog a human form, face down on a park bench, legs sprawled across the path. He knelt and examined what turned out to be an elderly man. To Norm's astonishment, it was the silent but unmistakable and thankfully still breathing figure . . . of his very own father.

Over a restorative cup of hot chocolate at a nearby coffee shop, while outside the fog still crept, the bearded man's remarkable story unfolded, of sudden loss of memory, of drifting vacantly to what he later discovered was France, where he woke in only a pair of shorts and sandals, without identity; of months wandering, picking up casual work as he could, avoiding the authorities for fear that he carried with him some dark secret about poisoned fish, of how he fell in with a gang of pickpockets who roamed Paris and only yesterday had come to try their criminal luck, as luck would have it, in Basingstoke, and how, as hazy recollections began to return to him in this his hometown, betrayal and lies saw him abandoned late in the evening, asleep on a park bench.

"I think," he said on the slow bus home, trying to understand his sudden state of bewilderment those two years ago, "it was the shock of our lottery win—£100,000 it was—and then my silly superstition about banks such that I chose to bring it home in cash in a shopping bag, then left it all on the number 63 bus.

"After so many years struggling to raise money for my old orphanage in Clacton, threatened by that developer, the loss was too much to bear. Out at sea, my poor, tortured soul and wounded mind simply chose to forget . . . everything."

"You never know, Father," said Norm, "it might turn up. Who knows, it could be lying undiscovered under this very seat," he laughed.

And Norm playfully reached down. There was indeed an abandoned plastic shopping bag wedged in the frame under the seat. Did they not clean buses nowadays? He pulled it out and peered inside at the usual lumpy detritus people were in the shameful habit of casting hither and thither, only to discover, amazed, great piles of tightly bundled £50 banknotes.

"Heavens above," said Norm's father, "that's it! I'd recognize the money anywhere."

"Excuse me, gentlemen," said an elderly female voice behind them. They turned to see a small, frail woman in a black shawl, peering strangely at Norm's father.

"I couldn't help noticing the unusual birthmark on your right ear. And then when you mentioned the orphanage at Clacton, a shiver ran through me. Forgive an old lady, but I've been searching for so long, you see, and I fear my mind plays tricks upon me. But I believe . . . no, one more question before I can be certain: Did you ever wear a silver chain bearing the small figure of a cat?"

"Ah," said Norm's father, "I'm afraid I did not. I perceive your hopes, but I am not the son you seek."

The old lady sank visibly. Another hopeful trail, perhaps the most hopeful of all, had led nowhere.

"But I know who is . . . "

She looked up. She smiled, hesitantly, a hopeful light restored to her eyes. She could scarcely contain the beating of her frail old heart. Could it be true?

"For there's someone else we must both now find together, an old friend, Bill, my greatest friend, a friendship first fashioned in our days in the orphanage, joined as we were in an improbable bond by identical birthmarks on our right ears. He always wore a silver chain exactly as you describe. He suffered a terrible breakdown in midlife at the memory of his abandonment in childhood and was confined thereafter to an institution. But alas, my own memory fails me now, and I cannot recall where. And yet for some reason I picture the sign of a ship in harbor."

"Like that one?" said Norm, looking out of the bus window at the very moment when, by chance, the fog began to lift, and pointing across the road to a sign above the door of a house of tranquil and fetching beauty.

"My God, that's it!" exclaimed Norm's father.

A few moments later Norm and his father were witness to the most tearful but blissful of reunions. At the sight of both his mother and his old friend, Bill recovered his wits in an instant. Moreover, the old nurse who had tended him with such devotion beyond her retirement age on account of the striking resemblance of a birthmark on his right ear to that of the son she, too, had been forced by cruel luck to give up to an orphanage in Clacton, was, of course, the mother of Norm's father, Norm's grandmother, only that day reeling from the news that the orphanage was finally to close for want of a last £100,000—and with it the last connection of many a mother to her long-lost child.

"Perhaps, Granny," said Norm, "this will help." And he held forth the shopping bag.

"You know," said Norm's father, "it reminds me of the time I was going around the Cape in the blackness of night in a force 9. The sea 'twas cruel, the ship's very hull did answer the wind in pain . . . "

. . . .

ONE DAY, on a cycling holiday stop in the Pyrenees, Mick Preston set off to the post office with a postcard for his friend Alan. On the way, who should he meet coming up the street, but Alan, on holiday. So he handed him the postcard. As Mick said: waste of a stamp.[1]

What coincidence? Oh, that coincidence. Fancy meeting the very person, etc. Well, if you say so. But how much of a surprise is it? Call it spooky, weird, whatever, but is it really so improbable that this should happen to someone, somewhere, at least once, and since it happened to Mick, is it a surprise that he talks about it? Events bump into each other all the time; it is only when people notice that we use the subjective word "coincidence."

In *The Art of Fiction*, the author and critic David Lodge says, "Coincidence, which surprises us in real life with symmetries we don't expect to find there, is all too obviously a structural device in fiction."[2]

If true, that says something odd about us. Because this chapter is full of coincidences and, Norm's story aside, they are all real. If we don't expect to find coincidences in real life, well, we should. There are plenty about. Though maybe not quite so many per head as Norm experienced one foggy day in Basingstoke. So is one reason that we often find coincidence clunking in fiction—as David Lodge says—because we make too much of it in life? Is coincidence overrated all round?

DS invited people to write to his website with their coincidence stories, a number of which follow.[3] He received thousands, all in one way fabulously unlikely. But if fabulously unlikely things happen so often, can they be that unlikely?

The problem—though it's also a huge evolutionary advantage—is that we are born seekers of cause and effect—as with the links between vaccinations and reactions—and we can't help trying to work out why things happen. "Nothing happens without a reason" is almost a description of how our basic cognitive kit functions. So if there is no particular reason, we're easily stumped. To call coincidences random is almost an insult to our normal sense of control and meaning.

But coincidences of the "fancy-meeting-you-here" type are certain, if not for you then for someone, somewhere, and they needn't mean anything. The opportunities are infinite, so why the fuss? There are enough of us with enough possible connections to make even the ingenious plot-driven inventions of Charles Dickens just about plausible. These coincidences are sometimes called artistic license. Do they need any? Or do they just reflect the fact that when lots of things happen, as things do, some are bound to be strangely coincidental?

But let's not strip all the romance from coincidences by making them trivial just yet. Mostly, so far, we have been dealing with the dark underbelly of risk—accident, death, disaster, gloom and doom—but coincidence can show the bright side of the play of chance in our lives. That's why coincidence has an important place in a book about danger: it is a version of danger's opposite—the turn-up.

What is a coincidence? It's been defined as a "surprising concurrence of events, perceived as meaningfully related, with no apparent causal connection." These concurrences may be two things that happen at exactly the same time: for example, a parent and child whose letters to each other cross in the post—after thirty-seven years without contact.[4]

Almost everyone seems to have a story of meeting a familiar figure in some unexpected place, or discovering some unexpected extra connection, such as the engaged couple who found they had been born in the same bed. Objects feature, too: such as buying a secondhand picture frame in Zurich, and finding in its lining a thirty-year-old newspaper cutting containing your own photograph as a child, or being on holiday in Portugal and finding a coat hanger that belonged to your brother forty years previously.[5]

Why do these extraordinary events happen? Various strange forces have been invoked, such as Paul Kammerer's principle of "seriality." As Arthur Koestler described it: "The central idea is that, side by side with the causality of classic physics, there exists a second basic principle in the universe which tends towards unity; a force of attraction comparable to universal gravity."[6] Kammerer says seriality is a physical force, but he dismisses as superstition any supernatural ideas that could, for example, link dreams to future events. In contrast, the psychoanalyst Carl Jung reveled in paranormal ideas such as telepathy, collective unconscious, and extrasensory perception, coining the term "synchronicity" as a kind of mystical "a-causal connecting principle" that explains not only physical coincidences but also premonitions.

Alas, more mundane explanations are possible.[7] First, some kind of hidden cause or common factor could be present—maybe you both heard that the Pyrenees were a nice holiday spot. Psychological studies have identified our unconscious capacity for heightened perception of recently heard words or phrases, so that we notice when something on our mind immediately comes up in a song on the radio. And of course, we only hear about the matches that occur, and not of all the people you have spoken to with whom you had nothing in common and indeed were pleased to escape. Few feel excited enough by not meeting a friend in the Pyrenees to tell anyone about their non-event.

Even so, people find coincidences weird. Let's try another way of making them ordinary—by making the extraordinary common. When DS tried his luck on Britain's National Lottery with the numbers 2, 12, 15, 25, 32, and 47, he lost. The winning numbers that came up that week were 4, 15, 19, 44, 45, and 49.

Extraordinary!

"What's extraordinary about that?" you say. He lost. Yes, but what are the chances of his six and the winning six being in this combination of 12 numbers? This is an amazingly rare combination of circumstances with a probability of 1 in 200,000,000,000,000 (1 in 200 trillion), the same chance of flipping a fair coin 48 times and it coming up heads every time. Impressed?

No, you're not impressed, we can tell. But why not? Well, you say, because the lottery had to come up with some set of numbers and so, too, did DS, and what's clever about them being different? But that misses the point. The chance of that precise combination—or of any other—was vanishingly small, *and the same as any other*. And yet it happened, as some combination has to happen. Only after the event does it seem boring, and only then because we give particular meaning to it. But the numbers don't care about the meanings we attach to them. They just come up with the same massive degree of unlikeliness in every case. But again the question, if everything is equally unlikely . . .

So it's purely the problem of predictability by people that gives this week's lottery numbers any meaning. The probabilities, or rather the improbabilities, remain the same for all combinations of guesses and results.

The chance of winning the jackpot on Britain's National Lottery is about 1 in 14 million, and in the Powerball lottery in the United States it is about 1 in 175 million. The only thing you can do to improve your prospects is try to pick numbers that other people won't, so you're less likely to have to share your winnings. So, since many people pick birthdates, avoid numbers below 31 (no one is born on the 32nd of the month).

Something similar applies to coincidences. Predicting precisely which coincidence will happen would be next to impossible. But afterward: so some numbers/coincidences came up? And? . . . Some combination of events is inevitable. Things have to happen. Someone wins the lottery. Maybe what happens is not meeting a friend in the Pyrenees, but maybe the Alps, or maybe not a friend but a relative, or maybe not carrying a postcard but having thought of them at breakfast, or maybe while singing their favorite song, or maybe not on the path but in the bar, or maybe not to this person but to someone else, or maybe . . . suggesting that there are innumerable potential coincidences and presumably innumerable near-misses. And whatever it is that happens to happen among all these possibilities that also turns out to be memorable, that's what happens to be talked about.

Coincidences are just an excuse to pretend that big numbers with limitless possibilities are meaningful in the small and everyday compass of our own lives. They're not. They just happen to catch our attention. They are a human vanity.

But tell that to the passenger who traveled by coach from Limerick in Ireland to London and took a copy of *The Name of the Rose* to pass the time, then left the book, by accident, unfinished, on the bus on arrival in London. On the return trip about three months later, there in the pouch at the back of the seat, was a different copy (different cover) of the same book.

Simple chance can be a strange and unintuitive force that throws up surprising concurrences more often than we might think, since truly random events also tend to cluster. Just as, if you throw a bucket of balls on the floor, they won't arrange themselves in a regular pattern, but will probably cluster here and there, so people moving randomly will sometimes find themselves gathered in one place. Ever been sitting on the subway to look up and find all the passengers on one side?

This characteristic produces brain-mangling results. For example, it famously takes only 23 people in a room to make it more likely than not for two to have the same birthday. We think the best quick way to see this—for those who find it intuitively outrageous—is simply to get a feel for the number of possible pairs of two people from 23 people. Say at the end of a game of soccer every one of the 23 people on the field (two teams and a referee) had to shake hands with everyone else. There would be 253 handshakes (and thus 253 possible birthday pairs). That is, there are an awful lot of possible combinations of two from among 23 people.

This, of course, means that in around half of all soccer games there will be two people on the field who share a birthday. Maybe they could give each other a hug. And that's just birthdays. Now let's think of everything else that any two players or the ref might have in common.

Given all the possible places and all the possible ways that two acquainted people could meet, or all the possible things that apparent strangers might have in common, once you start multiplying the possibilities you wonder if every day on the seats of the New York subway or slumped on a park bench in the fog, or tucked in a discarded shopping bag, there sit more potential connections than particles in the universe. Some will be a match, somehow, and never know it. Think of Dickens as choosing his plots from an infinity of real-life events, after the event: perfect realism. Or imagine that Norm was real, and he passed the audition to be in this book only because he had experienced this remarkable set of coincidences. Otherwise, we'd have chosen someone else.

The final explanation for coincidences is what is called the "law of truly large numbers," one version of which says that anything remotely possible will, if we wait long enough, eventually happen. So even genuinely rare events will occur, given enough possibilities. Take three children in a family. The first comes along on a certain day. The chance of the next one sharing the same birthday is 1 in 365, multiplied by another 1 in 365 chance for the one after that, which equals a 1 in 135,000 chance of them all sharing the same birthday, and better if there is planning going on. This is rare. But there are a million families in the United Kingdom with three children under 18, and according to the US Census Bureau, in the United States there were 4,917,000 households with three of their own children under 18 in 2011. So we should expect around 8 of the UK families to have children with matching birthdays, with new cases cropping up around once a year, which they do: new examples in the United Kingdom occurred on January 29, 2008; February 5, 2010; and October 7, 2010. And these are only the ones that got in the papers. In the United States, we would expect to find about 36 three-birthday matches.[8]

It would be truly strange if memorable coincidences didn't happen to you. But this may be difficult to keep in mind when you're walking past a phone box, it rings, you decide to answer it, and you find the call is for you.[9]

It's the detail that makes us sit up. "A postcard, you say? Fancy that. In the Pyrenees? Well, of all places." But the more detail we allow, the more possible points of contact there are. Just think of all the people you have ever known. Then think of all the people that you have had some connection with, such as attending the same school, being friends of friends or family, work colleagues, colleagues of colleagues ("Did you by any chance know of a guy named? . . ."), and so on. There will be tens of thousands. Perhaps the best way to think about coincidences is not how rare and strange they are, but how many we might be missing. If you're the sort of person who talks to strangers, you will find more of these connections. If you're not, then you have probably sat on a train next to a long-lost twin from whom you were separated at birth and never realized it. Coincidences are simply a refreshing reminder that we live among infinite possibilities. Even as that thought makes them less surprising, it also makes them wonderful.

That might never free coincidences in fiction from being obvious structural devices, but perhaps we should go easier on them. The only problem is, if we did lose some of our wonder at coincidence, would we lose some of our fascination for the story?

8

SEX

Kelvin's diary, aged 19¾:

Wake up. Head hurts.

Kath in the bed—asleep. "Oi, Kath!" Have shag.

Read paper. Seems familiar.

Last week's paper.

Open curtains. Sunny day.

Close curtains.

Half can Bud under bedside table. Down Bud. Have shag.

Sleep.

Wake. Hungry.

3 p.m. Find socks. Off to Mr. Singh's.

Buy lunch for two: Hershey bar for yours truly, Hershey bar for
     Kath, can of Stella, can of Bud.

Eat Hershey bar. Eat Kath's Hershey bar. Down Stella.

Home. Hi Kath. Put Bud in freezer.

Kath says where's lunch, d***head. Makes casserole.

Put casserole on stove.

Have shag (sofa).

Have nap (sofa).

Get up.

Strange smell. Brush teeth.

Strange smell cont'd.

Remember casserole.

Notice "do not put on stove" instruction on casserole dish.

Ignore "do not put on stove" instruction on casserole dish.

Remember can of Bud.

Open freezer. Can of Bud explodes.

Can of B remnants moving very fast.

Observe flight path of CofB pieces.

Duck.

CofB piece hits cupboard.

Kath not ducked.

Kath struck on shoulder by flying remnants of CofB.

Kath hits floor.

Casserole explodes.

Kath in contents of casserole shower.

Kath lying down, moaning.

Can see her panties. Red. Looks nice.

Suggest shag.

No shag.

Emergency Room crowded.

10 p.m. Home. No Kath. Kath's parents collected Kath, Kath fractured clavicle. Kath multiple burns.

Scrape casserole off stove. Nice.

Call Emma.

Be honest, what grabs you: the number 5.6 per 100,000 population, or the fragments of an exploding can of Bud pinging round the kitchen? The first is the rate of new cases of syphilis diagnosed in the United Kingdom in 2011. The can of Bud is an improbable image, no more, a crazy metaphor for carelessness. But it's the image that sticks in the mind.

Visual images of danger are usually sexier than numbers, for obvious reasons: they smack the senses with sound, color, movement, and violence. More sneakily, they also slant the way the danger appears, showing us its consequences, often ignoring the probabilities.

It's the same split between consequence and probability that we met in the chapter on accidents, when Norm fixated on losing his willy to a pike. It's not likely, but it's everything. Pictures of danger are usually the same, a vivid image of a bad "what if?"* and not a probability. The danger of skiing—which, based on the odds, ought to be illustrated by someone simply skiing safely, having (dare we say it?) fun—is more likely to be pictured as a leg in plaster or a whirl of skis and snow off a precipice. We don't "see" odds— how likely the thing is—we "see" consequences. That's what people would mean if they were to say "picture the risk." They mean picture the worst that can happen.

Advertisers and governments know this. When they want to change our behavior because "it's risky," or persuade us to buy something to make us safer, they often choose images of worst-case "what ifs?"

One of the most famous sex-risk images in the United Kingdom was what's known as the AIDS monolith advertisement in 1987. As the smoke clears, we see something like a tombstone carved in rock blasted from the mountainside. The chilling, brilliant voiceover is by the actor John Hurt:

> There is now a danger that has become a threat to us all.
> It is a deadly disease and there is no known cure.
> The virus can be passed during sexual intercourse with an infected person. Anyone can get it, man or woman.
> So far it's been confined to small groups, but it's spreading. . . . If you ignore AIDS, it could be the death of you.

Other public information films about danger, from a time when British governments still made them, include one about road safety with an image of a hammer swinging into a peach:

---

* The fashion for data visualization is an exception, where pictures of the numbers can be striking. See, for example, David McCandless's *Information Is Beautiful* (Glasgow: Collins, 2012).

It can happen anywhere, to anyone.

An ordinary street.

A moment's thoughtlessness.

Then splat, the hammer hits the peach.

Another, on the risk of crime, has hyenas prowling around a parked car. All very vivid and scary—and all consequential, not probabilistic.* But then, how do you do an image of one in a thousand, or any other odds? Not easily, not vividly. Numbers don't splat like the flesh of a peach or a human being; they don't bite like hyenas. In contrast, images of the worst-case "what ifs?" are easy, especially for one victim. And then comes the extrapolation to everyone else, as we're told that this tombstone, victim of hyenas, or splatted peach could be you. Think again of Norm's paralysis as he imagines the teeth and glassy eye of the pike in the reservoir. The consequential image dominates the odds for him, too.

Better than public information films and their selective images—don't you think—is private information. "I know from personal experience . . . " is the kind of know-all phrase we all love to use (and hate to hear), and certainly we grant them more authority than we do governments. But personal experience and sex? Authoritative and rigorously unbiased? Not selective? Yeah, sure.

Let's say you have unprotected sex, like Kelvin and Kath. If nothing goes wrong, you feel relieved. But you might also feel the risk probably wasn't that high after all. Exposure to danger can have that effect—of making people feel safer. "See, it was okay!" One formal explanation for this is that small samples of rare events are skewed. This means that when bad results are not typical—say only 1 time in every 20—you might get away with the first few, so you expect

---

* On the UK National Archives you can watch the AIDS monolith film, the peach and hammer (from 1976), and the hyenas. See National Archives, United Kingdom, Public Information Films: AIDS Monolith, www.nationalarchives.gov. uk/films/1979to2006/filmpage_aids.htm; Peach and Hammer, www.national archives.gov.uk/films/1964to1979/filmpage_hammer.htm; Crime Prevention— Hyenas, www.nationalarchives.gov.uk/films/1979to2006/filmpage_crime.htm.

to be fine next time, too. Your chance of being all right is 19 out of 20, or 95 percent.

You shouldn't deduce anything about the true odds from one experience. But deduce you will. Based on your own massively selective sample of the past, you might learn to underestimate probabilities in the future. "Had shag, no kid," as Kelvin might say. Do this a few times and he will—probably—still be okay, encouraging him to think—as if he needs encouragement—that this is the sort of risk he can take.

On the other hand, if it all goes wrong, people tend to overestimate the risk that it will go wrong again. Suppose unsafe sex leads to a sexually transmitted disease (STD) about 1 time in 50. It doesn't, but bear with us. People might not know the figure, even roughly. After 100 people have 5 encounters each, roughly 10 have an STD.

These 10 might say, "Crikey, we did it only five times and look what happened." So they take away an exaggerated sense of the odds. The other 90 might imagine that the probability of danger is not much above zero: "Well, it never happened to me." There are lab experiments—not involving sex—that seem to confirm this pattern of belief and behavior.

So personal experience can lead us astray. Again, we need bigger data than just our own notches on the bedpost.

Sex, like so many other fun activities, clearly carries dangers. These range from pregnancy (if you are a woman having sex with a man and don't want to be pregnant) to potentially dangerous disease or mild irritation, heart attack, injury when the bed collapses, hurt feelings—not even a phone call?—or the chance of embarrassment or arrest if you are caught having sex in a public place.

Let's start, as most of us did, with simple, unprotected sex between a man and a woman. What is the chance that any one of Kelvin's romantic encounters will end in a pregnancy? This, too, for understandable reasons, is difficult to study in laboratory conditions. A New Zealand study in which participants were only allowed to have sex once a month suffered, unsurprisingly, from a high dropout rate. Perhaps the closest has been a European study that recruited

782 young couples who did not use artificial contraception and who carefully recorded the date of every act (and there were a lot of them) until there had been 487 pregnancies.[1] The time of ovulation was estimated for each menstrual cycle.*

The bottom line is that a single act of intercourse between a young couple has on average a 1 in 20 chance of pregnancy—this assumes the opportunity presents itself on a random day, as these things sometimes do when you're young.

People who study how populations change are called demographers, and they use the term "fecundability" for the probability of becoming pregnant during one menstrual cycle. This varies among couples, but the average is estimated to be between 15 and 30 percent in high-income countries.

We can use the lower figure to illustrate the consequences for an average couple trying to have a child: there is an 85 percent chance of finishing each month without conceiving, and if we assume each month is the same and independent, then there is a $0.85 \times 0.85 \times \ldots \times 0.85$ (12 times) chance of not getting pregnant in a year of trying, which is 14 percent. So fecundability of 15 percent means a $100 - 14 = 86$ percent chance of getting pregnant in a year: a figure of 90 percent is often quoted as the proportion of young couples who will be expecting a child after a year without contraception, which corresponds to a fecundability figure of 18 percent.

Fecundability can be estimated from large populations if there is no effective contraception. In Europe this means going into history:

---

* The simplest way to estimate the chance of pregnancy was by considering only cycles in which there had been only one act of intercourse: the peak chance was observed at three days before ovulation, in which 8 out of 29 (29 percent) "coitions" resulted in pregnancy. But this only used around a third of cycles in which there had been any sex, and so a more sophisticated mathematical model was applied to the full data: the researchers estimated the peak at two days before ovulation, with a chance of around 25 percent, which is similar to previous estimates. The chances drop fairly steeply away from this peak, with an average of 5 percent over the whole monthly cycle. B. Colombo and G. Masarotto, "Daily Fecundability," *Demographic Research*, September 6, 2000, 3, www.demographic-research.org/Volumes/Vol3/5/default.htm.

more than 100,000 births taken from registers in France between 1670 and 1830 have been analyzed to produce an estimate of average monthly fecundability of about 23 percent.[2]

But suppose you don't actually want to get pregnant—How much is fecundability reduced by different types of contraception? This is generally expressed as the pregnancy rate following one year of use and strongly depends, of course, on how careful you are.

Contraceptive pills, intrauterine devices, implants, and injections are quoted as 99 percent effective, so that less than 1 in 100 users should be pregnant after a year, while male condoms are around 98 percent effective if used correctly, and diaphragms and caps with spermicide are said to be 92–96 percent effective, so between 4 and 8 women using them will be pregnant after one year.[3] Imperfect, or what's often referred to as "typical," use of contraceptives—you

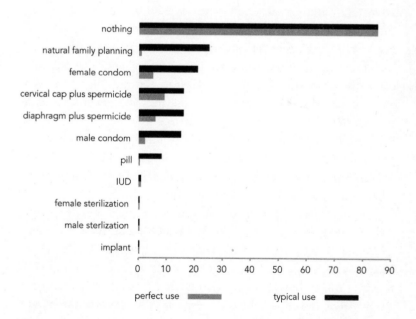

FIGURE 11. Percent of women becoming pregnant in one year using selected forms of contraception.

threw up the pill in the gutter on Friday night, forgot, it fell off, etc.—is, unsurprisingly, a lot less effective.

One way to make these comparisons is to imagine a woman considering various contraceptives. Assume an infinitely long life and sex life. How long before she typically becomes pregnant? In the case of female sterilization, she would expect one pregnancy in about 200 years. Similarly, if 200 women were sterilized, 1 of them would be expected to become pregnant within the next year. The standard way of expressing the risk is to say that it is 1 in every 200 "woman-years." For contraceptive implants the failure rate is so low that an accurate figure is hard to calculate. By one estimate, it is about 1 pregnancy every 2,000 woman-years.

If they don't have access to contraception, then women can end up having a lot of children. The French data showed that women in the 1700s who married between the ages of 20 and 24 had, on average, 7 children each, as women still do in Niger and Uganda. "Total Fertility Rate" is the average number of children expected per woman if current fertility carried on through her life. It took the United Kingdom more than 200 years to reduce the number of births per woman from 5.4 in 1790 to 1.9 in 2010, while other countries have shown a similar decline in just a generation: Bangladesh took only 30 years to go from 6.4 in 1980 to 2.2 in 2010.* Some countries have a remarkably low fertility rate, particularly prosperous countries in Southeast Asia (Singapore's is only 1.1) and countries in Eastern Europe—the Czech Republic's is 1.5, well below what is necessary to replace the population.

Pregnancy is not generally considered a great idea for young girls. In 1998 in England and Wales, 41,000 girls between the ages of 15 and 17 conceived—that's 47 in every 1,000, or 1 in every 21. Imagine that at school. The UK government set a target to halve this rate by 2010, and by 2011 the headline figure had dropped to 31 per 1,000 girls, a huge fall, which was especially steep in the three years

---

* Hans Rosling depicts the trends vividly at Gapminder, "Children Per Woman Updated," 2012, www.gapminder.org/data-blog/children-per-woman-updated/.

up to 2011. There is wide variation in these rates around the country—from 9 per 1,000 in Waverley to about 60 in Blackpool: that's 1 in every 16 girls between the ages of 15 and 17, pregnant every year. There is a strong correlation with low educational attainment, and deprived seaside towns like Blackpool traditionally have high rates. Great Yarmouth's is about 40 per 1,000.[4]

American teenagers are going the same way. In fact, the decline in the United States started earlier—more than two decades ago—with the pregnancy rate for girls between the ages of 15 and 17 dropping by almost half from 1990 (77.1 per 1,000) to 2008 (39.5). This is still higher than in the United Kingdom—and a lot of other places, too—but the gap is closing.[5]

Nearly half of these teenage pregnancies, in both the United Kingdom and the United States, end in abortions, but there are still many births: A 2008 report from the Organisation for Economic Co-operation and Development (OECD) put Mexico at the top, with 64 births per 1,000 women between the ages of 15 and 19; the United States fifth, at 35; and the United Kingdom seventh, at 24, while at the other end are Japan and Switzerland at fractions of the others, 4.8 and 4.3, respectively.[6]

But are babies a risk? It depends. We could turn these dangerous odds around and make them the measure of hope for those who want one. Pregnancy isn't always best described as if it were like the clap. It can be a blessing.

Which you just can't say about disease and infection. HIV, syphilis, gonorrhea, chlamydia, hepatitis, and numerous other STDs—some potentially fatal, some unpleasant—are dangers of unprotected sex of all sorts, and some even with protection, whenever there is a chance your partner is infected.

Risks are higher for young adults, men who have sex with men, people who inject drugs, and black African and black Caribbean men and women, who continue to be disproportionately infected. The peak risk in women tends to be among those 19 or 20 years old, and in men a few years later.

There has been an increase in the number of diagnoses for

many STDs in the past decade, partly because people's behavior has changed, mostly because we are simply testing a lot more people through screening programs and using more sensitive tests.

But the longer-run data are fascinating. Figures 12 and 13 reveal a rich history of sometimes scandalous behavior linked to world events—wars, social shifts in attitude, AIDS, and medical discovery—all expressed through sexual risk. Just look at the 1940s in Figure 12—the chart on syphilis—and draw your own conclusions (noting that penicillin became widely available soon after). The rates of syphilis during the war correspond to about half a million cases a year. They dwindled fast in the 1950s, 1960s, and 1970s, but then, in the late 1980s, briefly doubled, though "double" not very much is still not much, at least compared with wartime rates nearly ten times higher.

The slightly longer run of data available for gonorrhea in England and Wales shows more clearly the wartime spike in sexual disease, the large difference between men and women, and also the stark fall in cases that occurred alongside a sudden inhibition of sexual behavior in the 1980s that also coincided with the terror of HIV-AIDS.

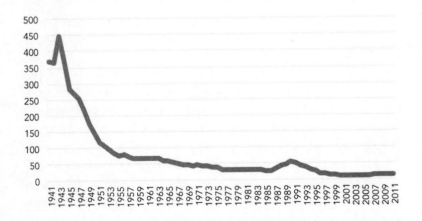

FIGURE 12. Rates of all-stage syphilis in the United States per 100,000 population.

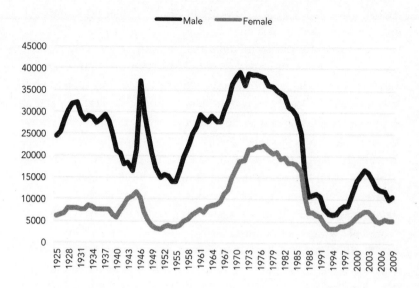

FIGURE 13. Historical number of diagnoses of gonorrhea, England and Wales, genitourinary medicine (GUM) clinics.

The chance today that you'll catch something after one sexual encounter with an infected person is not a figure medical authorities like to throw around. It is extremely dependent on your partner's background, which disease they have, and what you do together.

Very roughly—and not to be relied on for personal guidance—the HIV risk for a woman with a man has been put at about 0.1 percent (meaning that if she has sex once each with 100 infected men, she will have a 1 in 10 chance of becoming infected herself); for a man with a woman it has been put at 0.05 percent, or 1 infection per 2,000 "events"; and for a man with a man it goes up to 1.7 percent, depending whether you are "insertive" or "receptive." Behind these very approximate averages there are as you would expect plenty of stories of people who seem untouchable and plenty more stories of once being enough.[7]

Again, these risks vary according to such factors as the strength of the virus in the infected person's bloodstream. For gonorrhea (see

Figure 13), the risk of infection has been reported as up to nearly 1 in 2 for heterosexual partners.[8]

And then we must remember that sex can be energetic, which itself carries a risk. It's been recently estimated that 1 in every 45 heart attacks is triggered by sexual activity. Luminaries such as Nelson Rockefeller, Errol Flynn, President Felix Faure of France, and at least two popes are said to have succumbed this way. Solo sexual activity used to be associated with blindness and stunted growth, for which let's say there is a limited evidence base, but if it involves asphyxiation it is not recommended for the cautious. Numerous fatalities have been recorded or suspected, including David Carradine, Michael Hutchence, and a British MP. A study recorded 117 deaths from just two states in Canada.[9]

Finally, sex has to be the moment to mention optimism. Kelvin isn't the sexually careful type (see the flying Bud metaphor/image). No surprises there. A variety of attitudes can complicate sex and risk, and he has plenty, along with a heroic lack of self-control (see eating Kath's chocolate).

The first of these attitudes is wish-fulfillment, behaving as if what you wish to be true, will be true, as in, "It'll all be all right." This is funny to watch in young children—"Can I drive the car, daddy? I won't crash."—less so in adults, but still common. The puzzle is that we know we do it, and we still do it. As the Buddy De-Sylva song "Wishing Will Make It So" (from the film *Love Affair*) says: "Wishing will make it so / Just keep on wishing and care will go." This also goes by the glorious academic name of "desired end-state" bias, in which you calculate that the risks are lower for no better reason than you would like them to be lower. "You won't get pregnant. It'll be fine." People are similarly reported to underestimate the chances of divorce or losing their jobs—because they don't want it to happen.

Call this positive thinking if you like, but bear in mind the following examples, useful because they quantify the "It'll be all right" tendency. First, the "demonstrated, systematic tendency for project appraisers to be overly optimistic" during building projects. What could possibly go wrong with a building project? Plenty, as we all know. Project appraisers know this too. But still they tend to underestimate the risks—so much so that the UK Treasury forces them to make a formal adjustment for over-optimism.[10] So for capital spending on developing new equipment, even when you think you've thought of everything that could reasonably go wrong, add between 10 percent and an astonishing 200 percent for all that probably will. For the time it takes, add between 10 percent and 54 percent. Even for standard buildings, add up to 24 percent to the projected bill; for nonstandard civil engineering, add up to 44 percent to your best estimate of the cost.

A related problem is known as the planning fallacy, famously demonstrated by 37 psychology students whose average estimate of the time it would take to finish their thesis was 33.9 days, or 48.6 days if everything went right off the rails. In the event, the average time was a whole week longer than that.

It's been suggested that we are "hard-wired for hope,"[11] since hope for a better future might encourage us to strive for it. If this is true, evolutionary fortune favors the optimistic. This is also a conjecture. For though it might be true that over-optimism helps Kelvin's genes spread far and wide, that's not what he's striving for. His is the wrong kind of optimism—optimism that he'll get away with it.

Another explanation for over-optimism is that you know yourself better than you know others. So when asked to think about the risk that someone else will be accidentally knocked up, and lacking stats about average behavior, we might try to picture a typical case, but really invent a crude caricature, give vent to a prejudice, maybe, simply because we know no better. Who is she, this slapper who typically gets pregnant? "Well, she's probably drunk, she gets around, she just doesn't care, she's almost asking for it." In other words, if we think about the risk of getting pregnant, we tend not to think of statistical averages, we tend once again to conjure up images of extreme cases. Then if we compare ourselves to these extremes, we might say: "That's who it happens to, and I'm not like that, so I'll be all right."

The moral of this? That given sufficient motivation—and sex often supplies it—risk is easily confused with morality, hope, and even convenience, especially in the stories we tell ourselves.

DRUGS

Prudence drained her cocktail of butanol, isoamyl alcohol, hexanol, phenyl ethanol, tannin, benzyl alcohol, caffeine, geraniol, quercetin, 3-galloyl epicatchin, and 3-galloyl epigallocatchin.

She felt the dark brown liquid flow around her mouth, sucked it down, swallowed, eagerly, gratefully. The chemical concoction seeped into her body, the liquid radiated warm, sedative comfort. She knew the effects; in her short life she had already come to depend on them. Her eyelids fell, she leaned back in her chair and let out a long, tranquilized sigh as her arm fell to the table with the empty cup.

"More tea, Pru?" said Norm.

* *

Kelvin had never felt so gray. Color gone, from him, from the parade of shops he dragged past, traffic that hurt his head to look at, and mad visions of his blood's redness leaching away.

Sensation gone. Except pain. Pain that began hours ago between his shoulder blades, spread through his ribs into a coat of ache with every step, metal taste in his mouth, insomnia-lethargy-nausea, dried-out irritability, anxiety, tearfulness, and shot concentration. He needed another hit, now, to ease the pain. But he was out. "Now" was all he could think of. Another, now.

Kelvin spent his entire youth seeking instant gratification. At eighteen, he'd done enough molly to ski in—bombing a half gram of powder at 5 a.m. at a festival one day, and after that he was hooked, the mad looking-and-not-seeing and then seeing it all weird in the streetlights. Loved it. Didn't stop until several years later, when his heart went different, pumped different, and he experienced twitching hands and feet, and aches. So he quit pure, powdered MDMA, took something for the aches, which became pains, took more, now craved that instead.

He knew where to go and knew the routine. Be there, just before 6, with the money. So here he was, coat of pain, sweating. He ducked inside. The man he knew only as "the man" was there, sure enough. Always the same man, same coat, same place, lurking in the back. They nodded. Kelvin walked over. The man had it. Kelvin handed him the cash.

Then the man said what he always said, the same taunt every time, knowing Kelvin would be back: "Shall I put the repeat prescription and the receipt in the bag for you, sir?" he said.

"Please," said Kelvin. And he left, cradling his Vicodin like a baby, fumbling at the foil inside the bag to pop some before he was even outside the pharmacy.

. . . .

"And you took it why?" said Prudence on another occasion.

"Ah, well, you see . . . balance of probabilities weighted by risk perception, . . . " said Norm.

"Oh God," she said.

Kelvin had been goading Norm again, this time saying Norm's idea of the wild side was two cups of tea.

"So the question was how to experience a risk sensation with modest objective risk," Norm said. And after refining the options— how to maximize frisson per MicroMort, etc.—it came down to horse-riding . . . or Ecstasy.

"Am I really hearing this?"

"Death is possible in either case, naturally, but the point is to stick to practical calculations on an objective basis."

"Is it really?"

"What else would it be?"

"And?"

"Do you see a horse? Serious adverse event every 350 exposures."

. . . .

PEOPLE HAVE TAKEN mood-altering substances ever since they had moods. Remains of opium poppy husks have been found in Neolithic settlements in Europe; natives of South America have long chewed coca leaves as a mild stimulant or to suppress pain; almost everything edible has been fermented to make alcohol.

There are many drugs, and about every drug's risks and benefits there are many opinions. In this chapter we'll sample that opinion, one of the widest, wildest, and most diverse spreads of opinion about any risk,* to see how diversity squares with numbers.

The arguments have been around long enough. Opium in the eighteenth and nineteenth centuries was variously described as an inspiration, a medicine, and the devil. Some writers romanticized it. Samuel Taylor Coleridge took laudanum (from Latin *laudare*, "to praise"), a tincture of 10 percent opium in alcohol, before writing the poem *Kubla Khan* (1797), "composed . . . in a sort of Reverie brought on by two grains of Opium taken to check a dysentery"— until famously interrupted by a person from Porlock.

By contrast, John Jasper in *The Mystery of Edwin Drood* (1870), by Charles Dickens, awakes in a squalid opium den alongside a drooling Lascar, a haggard woman, and a Chinaman convulsed by gods or devils. Jasper is a choirmaster, but in Dickens's hands opium betrays his degeneracy.[1] In Oscar Wilde's *The Picture of Dorian Gray* (1890),

---

* See, for example, David Nutt, *Drugs—Without the Hot Air: Minimising the Harms of Legal and Illegal Drugs* (Cambridge, UK: UIT Cambridge, 2012), for a vivid description of different views on illegal and legal drugs.

"There were opium dens where one could buy oblivion, dens of horror where the memory of old sins could be destroyed by the madness of sins that were new."

In politics and business, too, fashions changed, and not just toward prohibition or censure.[*] Drug production went from cottage industry to industry proper, beginning in about 1827 with Heinrich Merck's commercialization of morphine, an extract of opium, laying the foundation of the Merck pharmaceutical company. Britain, as now, had a trade imbalance with China, and whereas today Britain exports malt whisky to the Chinese, in the nineteenth century the East India Company fought two wars for the right to ship them opium.

Heroin, also known as diamorphine, was first derived from morphine in St. Mary's Medical School in London in 1874, but was rediscovered by the Bayer pharmaceutical company in 1897 and marketed as a nonaddictive painkiller and cough medicine. Meanwhile, cocaine was being extracted from coca leaves. "Cocaine toothache drops—Instantaneous cure!" was one product for children. Sigmund Freud was one of many cocaine enthusiasts, as was Sherlock Holmes, to Dr. Watson's disgust:

> "Which is it to-day?" I asked, "morphine or cocaine?" He raised his eyes languidly from the old black-letter volume which he had opened. "It is cocaine," he said, "a seven-per-cent solution. Would you care to try it?" "No, indeed," I answered, brusquely.[2]

Today a modern Holmes equivalent, Dr. Gregory House, in the TV medical drama *House*, can use and abuse Vicodin, a painkiller and opiate, just like Kelvin, even as politicians speak of a war on drugs. Romance and heroism still sit side-by-side with the end of civilization, amid the stars and in the gutter.

---

[*] A recent example of drug liberalization is the Licensing Act of 2003, which relaxed the rules on the opening hours of licensed premises, even allowing continuous opening for some premises, in England and Wales in 2005 (similar changes were introduced in Scotland in 2006).

Compare the following edited extract, from the oral testimony at a meeting of Narcotics Anonymous UK of a man who lost his brother to drug abuse, in which drugs are the mark of a life almost born to ruin, and who still says he is a sick man getting well—

All the men and the women I grew up around were involved in crime and violence. I didn't like the violence I was exposed to, me dad kicking fuck out of me and me mother and I'd be hiding me brother under the beds. And so I got out when I could. And what me dad done is he said I'd never be anything and he put all his hang-ups and all his negative perceptions and all his issues on me. And the way I coped with that is I became very paranoid and very angry with anyone who slighted me and I learnt how to be violent and how to get amongst people and hide really well. I started taking drugs when I was about 8. Robbing Diconal off the prostitutes. . . . Heroin in the area was an epidemic and I started selling. The expression is living off the land. And by the time I was 19 I ended up in a prison and got involved in a disturbance and ended up with a sentence of 19 years and 3 months. What I done is I used to live and I lived to use and whatever way I could get drugs in prison I done it. I just took [up] everyone's resentments . . . and I'd be walking round yards with these Palestinians and they'd be telling me what was going on in the Gaza strip and that and I'd say "can I be one of yous?" I had the full beard. And me mam's answer was: "I'd better bring him more gear, straighten him out" [laughter]. But underneath it all I was a broken young man with no identity.[3]

—with this, from Hunter S. Thompson, drug glutton and author of *Fear and Loathing in Las Vegas,* who had a journalistic style once described as extreme participatory fieldwork—

I like to just gobble the stuff right out in the street and see what happens, take my chances, just stomp on my own accelerator. It's like getting on a racing bike and all of a sudden you're doing 120 miles per hour into a curve that has sand all over it and you think "Holy Jesus, here we go," and you lay it over till the pegs hit the street and metal starts to spark. If you're good enough, you can pull it out, but sometimes you end up in the emergency room with some bastard in a white suit sewing your scalp back on.[4]

The point again is the variety of views—all the way from fatally corrupting to life-affirming. Ideas of what it is that's at risk have also shifted: it's not just the damage to health that people now talk about, but relationships, employment, crime, the economy, and the rain-forests cut down to farm coca—far away from the drug-taker.

To many, drug abuse is used needles, squalor, and everything the film Trainspotting is about. To others, it comes with a corporate logo. Some addictive drugs—painkillers like codeine—are available over the counter; others—alcohol in a bar—may be woven into a way of life. Cathryn Kemp, author of a memoir about addiction, "used to think a drug addict was someone who lived on the far edges of society. Wild-eyed, shaven-headed and living in a filthy squat," until she became one. What began as a legitimate need for pain relief and a painkiller called fentanyl became a comfort as she consumed it the way she'd take a glass of wine to unwind at the end of the day, then moved to dependency, then to about ten times the maximum rec-ommended dose, then to thinking only of the next hit, living only for the drugs, raging at her family, finally accepting that she was an addict, and undergoing a brutal rehab, all via nice packaging and clean suppliers at local pharmacies, much like Kelvin.[5]

The Times, the UK newspaper, reported in 2012 what it called the "Scandal of 1m Caught in Tranquilliser Addiction Trap," about "the failure of the NHS to provide meaningful support to victims of benzo-diazepine addiction, even though they significantly outnumber those hooked on illegal drugs." Unlike with heroin, there are no hard-hitting public information campaigns about how painkillers screw you up.[6]

Are Prudence, Norm, and Kelvin all users? They would disagree, as will readers. They would also differ over the severity of the risks. Which is worse: booze, Kelvin's painkillers, or cocaine? Does Pru's stimulant of choice (tea, but not fair-trade tea—so some will accuse her of a socially harmful addiction) belong on the same page as Norm's experiment with E, or is Ecstasy little more dangerous than a cuppa for many self-described recreational users who successfully hold down a job and relationship?

Knowledge changes, beliefs vary, attitudes are all over the shop, and numbers are only part of it. People's sense of harm is and always was plainly bound up with personal experience, anecdote, moral values, and personal preference—what's your poison?—as well as social norms, norms among friends, among their income, ethnic, and age groups, or in society as a whole, norms that are sometimes freely chosen and sometimes not: whether, for example, they turn into social pressures to conform, with a drink, or a tab.

Often the illicit drug-user's first line of defense is to say that attitudes to alcohol prove there's no consistent scale of harm, only shades of hypocrisy. In the decades since governments began to talk more about the dangers of drug use, two of the most potentially dangerous and addictive drugs—tobacco and alcohol—remained legal. And that's probably not about to change. As a result, it is said, "a whole generation learned to ridicule and ignore all governmental advice on the subject."[7] Others talk—not always in ironic tones—about "respectable drug addiction"—middle-class use of the opiate-based codeine to help with sleeplessness, for example. What could be less sinister than wanting a good night's sleep? But codeine dependency can be ugly and dangerous, too. Rules and attitudes said to be linked in large part to risk and harm have a fickle relationship with them.

Even so, and despite being keen on data, we argue that little of this mess is intrinsically wrong or irrational, although it may not always be honest or self-aware. It is simply that danger is only one part of a hideously complicated conversation. How do we describe the risks of drug use when they are part and parcel of what we value in life? Because this is a problem of weighing both personal harm and social harm (and pleasure), harm to a way of life, where sometimes that way of life is threatened by drugs—especially by the crime they bring—but sometimes may depend on them, such as when someone downs a few pints at the pub at Sunday lunch, takes half an E at a club, or smokes at the races. Add to that a fierce disagreement about how best to minimize the harm—by prohibition, decriminalization, or a war on drugs—and you have a recipe for muddle.

Questions of value give philosophers headaches. The point of saying all this is that risk can also fall—and should—into the same category of philosophy, one of competing values and traditions at least as important as the power of the data. So heated is the argument about illicit drug use that it is almost surprising anyone should think data about mortality or harm could settle it.

Perhaps the acid test (no pun intended) of the relevance of the data to this whirl or beliefs and emotions is how you react to Norm. What did you think when he tried to skip over all the values business with a purely data-driven approach to excitement, illegal or not? Is he the only sensible person on the planet? Prudence doesn't think so. Or is his kind of logic, in this instance, insane?

The anthropologist Mary Douglas argued that the assertion of risk—"don't do that, it's dangerous"—is often sly social control. If behavior offends us, for whatever reason, we warn that bad things will happen and call the behavior "risky." In her fieldwork she found tribal women who'd been told that if they were unfaithful they ran a greater risk of miscarriage, a risk presented to them as a fact of nature, a bit like the need to take care with a sharp knife. The supposed biological risks of the behavior were clearly a fiction; the true purpose, the social purpose, of warning about the risk of infidelity was evident: to control the women's behavior. Early in her career Douglas thought this the kind of thing only primitives do, since superseded by science. Later she argued that we all do it, in all cultures and all ages.

But wherever one person seeks control, another kicks back. Hunter S. Thompson couldn't consume enough illicit drugs, and he seemed more determined to stuff them down because others wanted to stop him. His accounts of his drug binges read like a knockdown fight with majority opinion. He describes his "twisted"—a good word, apparently—trip to Vegas as fulfillment of the American Dream. Some dream.

Thompson was not your everyday risk-taker. But was he unusual when he suggested that the best way to resist control is to assert control of your own, even unto being out of it? Transgression is partly what risk-taking is about, at least for some. Stepping over a boundary can be a thrill in itself. And we'd have all this reduced to a number?

So in front of any statistic about the quantified level of danger we need a flexible sign, positive or negative for different readers, depending on what they approve of, while also recognizing that the value they put on any risk isn't only up to them. Their behavior affects others. So calculating the dangers of drugs is both a personal judgment and a blazing social row replete with wider values and preferences at least as much as it is a statistical, evidence-based exercise.

Mind you, once sociology, anthropology, psychology, and the rest have said all this, you sure feel the need for some numbers—if not to make policy, then at least to help make up your own mind. What are they?

Concern about the addictive properties of opiates and cocaine in the early twentieth century led to criminalization for misuse. Criminalization makes it hard to estimate the level of harm from drugs, because you first need to know who uses them. If admitting this makes you a criminal, you might keep quiet.

So the British Crime Survey (BCS) guarantees anonymity when it asks about the use of illegal drugs. Responses are scaled up to the adult (ages 16–59) population of England and Wales, of whom around 1 in 3 (US: about 1 in 2) are estimated to have used illegal drugs in their lifetime, around 9 percent (US: 15 percent) in the past year.[8]

Figure 14 shows lifetime use for all Americans over age 12. Note that "lifetime" use tells us nothing about current use. Lifetime use can mean this morning or 50 years ago. For example, among 18- to 20-year-olds, around 40 percent of respondents say they have used an illicit drug in the past year, compared with a little under 10 percent for those 35 and older.[9]

What's so bad about drugs? They can certainly be dangerous in the wrong hands: in the United Kingdom, a Manchester general practitioner, Harold Shipman, injected over 200 of his patients with lethal doses of diamorphine (heroin) in a murderous career before he was finally caught in 1998, after a clumsy attempt to forge the will of one of his victims (see also Chapter 22, on crime). And there have been numerous famous deaths from drug misuse, whether

drug	% who have used in lifetime	% who used in past year
Marijuana and Hashish	42.8	12.1
Cocaine	14.5	1.8
Crack	3.5	0.4
All hallucinogens	14.6	1.7
LSD	9.1	0.4
Ecstasy	6.2	1
PCP	2.5	0.1
Inhalants	8.1	0.7
Heroin	1.8	0.3
Nonmedical Use of Psychotherapeutics	20.9	5.7
Pain Relievers	14.2	4.3
Tranquilizers	9.1	2
Stimulants	8.3	1
OxyContin®	2.5	0.6
Sedatives	3.1	0.2
Methamphetamine	4.7	0.4

FIGURE 14. Lifetime drug use in the United States.

deliberate or not—from Janis Joplin to the Singing Nun.

But working out exactly how many people die this way is tricky: in general, in the United Kingdom, if drugs are mentioned on the death certificate, it means the death is counted in the official data as drug related, even if the drug was not the sole cause.[10] Altogether there were 1,784 deaths in England and Wales in 2010 from misuse of illegal drugs (i.e., not including alcohol), down slightly from preceding years but double the figure in 1993.

The peak decade for men is the thirties, with 544 deaths in 2010. That's around 150 MicroMorts per year, 3 a week, averaged over everyone in that age group, from dopeheads to vicars. Almost exactly half the total deaths (791) were due to heroin or morphine. Cocaine was associated with 144 deaths, amphetamine 56, while those involving Ecstasy fell to only 8 after averaging around 50 a year from 2001 to 2008.

If we use the BCS estimates of the number of users, we get a rough idea of the annual risk, in MicroMorts, for users of different drugs. Averaging out from 2003 to 2007, cocaine and crack cocaine were involved in 169 deaths per year, and so an estimated 793,000 users were each exposed to an average of 213 MMs a year, around 4 a week.[11]

Ecstasy's 541,000 users experienced around 91 MMs a year each, or around 2 a week. Since the 2003 market for Ecstasy has been estimated as 4.6 metric tons (about 5 US short tons), this corresponds to around 14 million tablets, or an average of around 26 per user. This translates to roughly 3.5 MMs per tablet.[12]

Cannabis seldom leads directly to death, but its estimated 2.8 million users suffered an average of 16 associated deaths per year, which is 6 MMs a year.

This level of harm is as nothing compared with the average of 766 heroin-related deaths a year, which works out at 19,700 MMs per year—54 a day—like a 350-mile motorcycle ride every day in the United Kingdom, or the exposure *every day* to seven times the risk from a year's worth of cannabis, although this very high figure depends on the BCS underestimate of users, and so is an overestimate of the risk.

In the United States there were about 3,000 deaths connected to unintentional heroin overdose in 2012 and about 5,100 from cocaine. The numbers of users in the month before the survey was conducted were estimated by the US National Survey on Drug Use and Health to be 335,000 and 1,650,000, respectively, giving approximate risks of 9,000 MMs a year for heroin use (about 25 a day) and 3,000 a year (9 a day) for cocaine users. As ever, such comparisons must be made with caution: if you drive into a tree while high, does the record show death by road accident or by drugs? The United States tends to cite overdose or poisoning figures. The UK figures generally also include the tree.[13]

But the most striking trend in American drug use is that poisoning is now the leading cause of death from injury, ahead of motor vehicle traffic, and that 9 out of 10 of these poisoning deaths are caused by drugs. One class of drugs, above all, dominates this trend, and it is not necessarily illegal. Misuse or abuse of prescription drugs, including opioid analgesic pain relievers, accounts for a large proportion of the roughly sixfold increase in the rate of unintentional drug overdose deaths in the United States in the past three decades. The point we made earlier, that what people think of as stereotypical drug abuse isn't always consistent with a quantified measure or harm, is affirmed by the fact that legal drug users and illegal drug users became equal visitors to the ER in about 2007.[14]

The response to this trend will be interesting. If the physical harm caused by both begins to look more similar, and cultural attitudes toward one type are influenced by cultural attitudes toward the other, will this soften people toward illegal drugs, or harden them to legal ones? Will we see campaigns to urge people to "just say no" to painkillers? However this develops, what's the betting that it will be by quiet conversation? The relationship between fear, outrage, and data is on a long piece of elastic. That's not the same as saying that people's views lack reason, but simply that their reasons are not even nearly all about quantifiable risk, on anything like a generally agreed scale of harm.

Not least because there are many other harms than death that are caused by drugs: for example, it's been estimated that smokers of cannabis are about 2.6 times more likely to have a psychotic-like experience than nonsmokers.[15] Heroin injectors may contract HIV or hepatitis from nonsterile needles. They may have abscesses, or suffer poisoning from contaminants, apart from the risks of dependency and withdrawal, not forgetting the standard effect of opiates on bowel movement—the focus of an exchange in the play *A Voyage Round My Father*, where John Mortimer says to his father, "Did you ever smoke opium?" And his father replies, "Certainly not! Gives you constipation. Ever see a portrait of that rogue Coleridge? Green around the gills and a stranger to the lavatory."

So potential effects ripple out from bad guts to violence to felled rainforests and beyond. This is another way of stating the problem about how to make comparisons between harms that matter differently to different people.

Is it possible? It's been tried, let's put it that way, including an attempt to put a single, summary nastiness-number on each drug. A recent study that looked at harms of various drugs, including the legal ones alcohol and tobacco, took in such effects as mortality, damage to physical and mental health, dependency, and loss of resources and relationships as well as harms to society such as injury to others, crime, environmental damage, family adversities (such as harm to family relationships), international damage (like those rainforests), economic costs, and effects on the community. Each drug was scored on each dimension, the different harms weighted according to their judged importance, and a total harm score calculated. As with any composite indicator, what goes into the mix and the weight accorded any single factor is arguable.[16]

The drug at the top of the resulting ranking was alcohol, with a score of 72; then heroin and crack cocaine, at 55 and 54; tobacco was sixth, at 26; and Ecstasy almost at the bottom of the list, with 9, despite being a Class-A drug in the United Kingdom. This ranking was controversial, to say the least.

More controversial even than comparing illegal and legal drugs is to compare illegal drugs with wholesome activities. Professor David Nutt, then head of the UK Advisory Council for the Misuse of Drugs (ACMD), wrote a paper comparing Ecstasy with "equasy," the addiction to horse-riding, claiming that both were voluntary leisure activities for young people of comparable danger.[17] He did not remain chairman of the ACMD for long. Not because his numbers were absurd. They weren't. But risk, to repeat ourselves, is not simply, or even mainly, about the threat of harm. His political bosses seem to have decided that comparing equestrian activities to drugs wasn't good politics.

Professor Nutt said the dangers of equasy were revealed to him by the clinical referral of a woman in her early thirties who had suffered permanent brain damage as a result of equasy, leading to severe personality change, anxiety, irritability, and impulsive behavior, bad relationships, and an unwanted pregnancy. She lost the ability to experience pleasure and was unlikely ever to work again.

He wrote:

What is equasy? It is an addiction that produces the release of adrenaline and endorphins and which is used by many millions of people in the UK including children and young people. The harmful consequences are well established—about 10 people a year die of it and many more suffer permanent neurological damage as had my patient. It has been estimated that there is a serious adverse event every 350 exposures and these are unpredictable, though more likely in experienced users who take more risks. It is also associated with over 100 road traffic accidents per year. . . . Dependence, as defined by the need to continue to use, has been accepted by the courts in divorce settlements. Based on these harms, it seems likely that the ACMD would recommend control under the Misuse of Drugs Act perhaps as a class A drug given it appears more harmful than ecstasy.[18]

He concluded that a more rational assessment of relative drug harms was possible, but the drug debate took place without reference to other causes of harm in society, which gave drugs a different, more worrying, status.

Is that status deserved? If we think so, we need to say why, without saying that it is because drugs are more harmful, since harm alone doesn't bear the weight of opinion. Quantifiable risk is not the answer.

The final warning comes from Dr. Watson as he witnesses his friend Sherlock Holmes indulging in a (then perfectly legal) drug in *The Sign of Four*:

"But consider!" I said, earnestly. "Count the cost! Your brain may, as you say, be roused and excited, but it is a pathological and morbid process, which involves increased tissue-change and may at last leave a permanent weakness. You know, too, what a black reaction comes upon you. Surely the game is hardly worth the candle. Why should you, for a mere passing pleasure, risk the loss of those great powers with which you have been endowed?"

10

BIG RISKS

Kelvin peeled away down a steep path, marched to the door, scuffed to a stop, and knocked. Norm scampered after. Kelvin had a low opinion of people who disagreed with him. Dodgy motives the lot. Climate change was an excuse for wacko eco-zealots to tell you what to do. Activists were control freaks, and Greenpeace gave him Tourette's. As for all that stuff about polar bears . . . blackmail with a furry animal.

"And whales should be used up," Kelvin said. "Why not? We need soap."

A woman in a wide skirt, hazy through frosted glass, swayed down the hall, the latch clicked, the door half-opened, and through the gap came a cautious smile with a gray bob, somewhere in her late sixties.

He and Norm were in one of those bad first jobs that people do desperate for a first job, trudging door to door, hating suburbia, failing to drum up business and putting the world to rights instead.

Norm had dared to suggest that majority opinion accepted the case for anthropogenic climate change and Kelvin should consider the balance of probabilities, but Kelvin said that only showed people were f——wits, and to prove it . . .

"Good afternoon, my name is Mr. Poe," said Kelvin, offering his hand through the gap. "And this is my associate, Mr. Edgar."

"Oh, yes?"

"From London Zoo."

"Oh?"

"You'll have seen the program—on the television last night."

"Which one was that?"

"About the zoo. The penguins."

"I watch the BBC. For *Eastenders*."

"Of course. But you'll know, you'll have heard about the penguins."

"Well, I don't know. What about the penguins?"

"Being closed."

"Closed?"

"That's why we're here. Closed down. Climate. Habitat. It's an absolute urban heat sink in a city zoo, you know."

"Oh. What will they do?"

"Quite. That's why we hope you can help. That's what the appeal is about."

"An appeal, I see."

"You're most understanding," said Kelvin, who advanced like a vicar to the needy and took her hands. "And you do have a nice bit of garden at the back there, high up, flood-free."

"Oh yes," she said, letting go of his hand.

"There you are then. Just right, isn't it, the garden?"

"Well, I like it."

"Excellent. I'll put you down for two."

"Two?"

"In the morning."

"I'm sorry, I'm not buying . . . "

"No, no, they're free, completely free, though in the circumstances you're the generous one."

"Well I, I don't . . . "

"No need to worry, full instructions with every penguin. About eight o'clock all right?"

"Well, no . . . "

"Two penguins to number 17, Mr. Edgar."

"What?"

"Penguins, Mr. Edgar."

"There's been a misunderstanding. I'm sorry . . . "

"Plenty of shade," said Kelvin turning away. "And they like muesli."

"But I eat cornflakes."

"Cornflakes will do. But not too much or they get fat. Come, Mr. Edgar, sixty-four to go."

With a nod: "You're so kind. The penguins will be most happy here."

He marched.

"No, you can't . . . "

Up the path, Norm trotting after.

"Really. This can't . . . excuse me . . . "

Brisk. "You!" . . . and "You!" . . . gone.

. . . .

At 08:39 GMT the following morning Michael Spiegelhalter of the Catalina Sky Survey made observations using the Mount Lemmon 1.5-meter aperture telescope near Tucson, Arizona.

He was puzzled. The potential near-Earth object SO43 didn't seem to be quite where he expected. An error, probably. He emailed the Minor Planet Center in Cambridge, Massachusetts, to let them know they might like to check the data.

. . . .

WHAT DOES SUCKERING an old lady about penguins have to do with big risks like climate change?* Plenty. But to see it, we have to step outside our own beliefs about these risks. So, just for a moment, forget whatever you think is the truth about climate change and accept that, as risks go, opinion touches the extremes—from nothing much

---

* By "big risks" we mean those that affect lots of people, such as the climate, new diseases, natural disasters, and so on.

to worry about, to the end of life as we know it. Is global warming a hoax, as Kelvin thinks? Or are we going to fry? People disagree.*

One side says the great majority of scientists who study climate change think the planet is genuinely warming, that people are responsible for it, and it's serious. The skeptics on the other side are wrong, conspiracy theorists, nutters in the pay of . . . etc.

Skeptics like Kelvin appeal to science, too, and return the abuse. The great majority of proper, competent, honest scientists think human-made climate change is hot air, they say. The rest are conspiracy theorists, nutters, in the pay of . . . etc.

They say this about even the most basic question of the lot: Is the world getting hotter? Forget whether people are responsible for it, or what to do about it, just that simple, measurable detail: Is the world getting hotter? Still people disagree. Still they think the science—the real science—is on their side.

Which is not to argue that climate change is all a matter of opinion. But it does show how opinion about danger can work wonders with much the same evidence. So what drives the views people take about what ought to be—in an argument where everyone says evidence is paramount—basic scientific facts?

We could say the answer is trivial and obvious: that everyone is biased. People see what they want to see, scientific facts included. The beginning of a more interesting answer is that the facts as we see them about big risks like climate often depend less on the facts

---

* The Guardian reported that in the United Kingdom "thirty one per cent said climate change was 'definitely' happening, while 29% said 'it's looking like it could be a reality,' and another 31% said the problem was exaggerated, a category which rose by 50% compared to a year ago. Only 6% said climate change was not happening at all, and 3% said they did not know." Juliette Jowit, "Sharp Decline in Public's Belief in Climate Threat, British Poll Reveals," The Guardian, February 23, 2010, www.guardian.co.uk/environment/2010/feb/23/british-public-belief-climate -poll. About a year later, the paper reported that a new poll had confirmed this trend, "with just 14 percent saying global warming poses no threat." Damian Carrington, "Public Belief in Climate Change Weathers Storm, Poll Shows," The Guardian, January 31, 2011, www.theguardian.com/environment/2011/jan/31/public-belief-climate -change.

than on who we are. Kelvin's politics, love of freedom, irritation with government, tendency to risk-taking and impulsiveness, even his loathing of the conformist wasteland of suburbia, are not co-incidentally linked to his beliefs about the facts of climate change. They directly contribute to them. They also feed his scorn for people who can be taken in by the "lies" of the other side. So our views of the facts about big risks are often prompted by our politics and behavior, even as we insist that the rock on which we build our beliefs is scientific and objective, not the least bit personal, even as we swear we believe what we believe because it's true, why else? At least, that's the story we tell ourselves. It's everyone else who is bi-ased by personal and political baggage, and the story we tell of them is about their stupidity and corruption.

But it is not just the evidence offered by the other side that we dislike or disagree with; it's what those facts seem to tell us about the other side's whole political makeup. That's why Kelvin feels free to be rude to an old lady: if she's willing to believe all this penguin nonsense, she's probably a socialist.

We have already said, in the chapter on drugs, that values can drive people's sense of risk. Climate is in some ways similar, but a tougher case to prove, since few people are willing to admit that their views on the risks of climate change, or even the basic fact about whether the world is getting hotter, arise from their political makeup. But that is the argument we will consider in this chapter.

A quick word of caution: if you think climate change is proof that cherry-picking scientific evidence is typical only of the politi-cal right, or left, hold fire. Those who study these habits say all sides have them, though the details vary depending on the risk they're talking about.

Scientists sometimes talk as if all that's needed to put right those they regard as misinformed is a little enlightenment with the facts. But if the argument in this chapter is right, their attempt to en-lighten them in this way will fail, and probably backfire.

Here's how the process works. Because the climate is a huge, complicated system, it is impossible to be certain what it's up to.

Persuaded, yes; certain, no. And the smallest doubt creates room for disagreement. This is where values creep in. To see how, let's take a quick digression via engineering at the micro-scale known as nanotechnology—"gray goo," to its critics. What do you know about nanotechnology? If the answer is anything like as much as MB and DS, not much: a bit of rumor, a bit of mystique, a lot of acronyms.

Next question: How do you *feel* about nanotechnology? Again, if you're like most people studied on this question, despite an absence of facts you will probably have a view: "excited" maybe, or "anxious." You may even have a few speculative reasons for the way you feel. Go ahead, be our guest, sound off, ignorance is no bar. Most of us do to some extent. We form our initial opinions about danger with the haziest grasp of the known facts.

Dan Kahan, at Yale University, leads what he calls the Cultural Cognition Project, studying how people form opinions about risk.[1] Kahan tells the story of the reaction in 2006 of the city authorities in Berkeley, California, to a proposal by the university for a nanotechnology research facility:

> Having never heard of nanotechnology before . . . the city's hazardous waste director immediately commenced an inquiry. . . . "We sent them a bunch of questions starting with: "What the heck is a nanoparticle?" Regulators were quickly able to learn that, but not much more: "The human health impacts of nanoparticles," the city's Environmental Advisory Commission reported, "are very complex and are only beginning to be understood." Nevertheless, citing concerns that nanoparticles might "penetrate skin and lung tissue" and possibly "block or interfere with essential reactions" inside human cells, officials concluded that a precautionary stance was in order.[2]

Knowing next to nothing, the city regulator still formed a view about the risks. He could decide, as he did—as can any of us—that things are risky until there's a hint otherwise. Or he could have decided that things are safe until there's reasonable evidence otherwise. So what made him jump one way rather than the other? Not the evidence, says Kahan, there wasn't much, either to suggest that nanoparticles were dangerous, or that they were safe. The deciding factor was his cultural disposition.

Funnily enough, whatever people's first instinct about risks like this, more information only seems to confirm it. Kahan and his associates went on to question another 1,800 Americans about nanotechnology, discovering that it was their visceral, emotional response, pro or con, that determined how risky they thought it was, and that these views generally only hardened the more people learned, suggesting that when we claim to take a provisional view, we're often kidding ourselves, and maybe others, that we're open-minded when our opinion is already a closing door, swung not by facts but a set of prior cultural attitudes.

Kahan's work suggests that people tend to assimilate new knowledge "in a manner that confirms their emotional and cultural predispositions." In other words, they filter the facts to suit their beliefs and instincts from the first. Belief doesn't simply follow fact, belief decides what the facts are. So, is nanotechnology dangerous? Ask instead: "How do you feel?"

Sometimes people say that they just don't know what to think, it's all too complicated. So we turn to the experts. But are we really surrendering judgment to our betters, bowing to their authority? Often not. That's because it's we who decide who qualifies in our eyes as an expert. And we pick experts who give answers consistent with the way of life and politics we happen to like. If you're a hippie with a wild beard and I'm a suit with a short haircut, I'm less likely to rate your qualifications, whatever they are. We judge the experts' expertise according to whether they seem to be people like us, people who share our cultural outlook. Again, Kahan's experiments seem to confirm this. He shows people pictures of made-up experts, including one with a wild beard and another with a suit and short haircut—all with impeccable academic credentials—to see what it is that people recognize as expertise. Sure enough, they choose their experts carefully. Social psychologists call this kind of selectivity "biased assimilation." Being the serious sort of person who reads books like this, you're above all that, of course. Except that Kahan finds that, the more scientifically literate people are, the more they do it.

He argues that we can all be placed on a spectrum of attitudes and beliefs,* and once he knows how you feel about nanotechnology, he has a good idea how you will feel about the risks of climate change, nuclear power, gun control, and so on. What's more, he says, you will believe, wherever you stand, that the true scientific consensus—not the nutters and conspiracy theorists—is with you. This is what Kahan means when he talks of "cultural cognition."[3]

One way of classifying people is to put them on a line that runs from "individualist" to "communitarian." Individualists tend to dismiss claims of environmental risk, "because acceptance of such claims implies the need to regulate markets, commerce, and other outlets for individual strivings," says Kahan, just as Kelvin doesn't believe the Green movement because he thinks that every time it sees a crisis it starts telling people what not to do.

---

* Kahan's work here draws heavily on the cultural theory of risk. One of the most influential accounts of this is in *Risk and Culture*, by Mary Douglas and Aaron Wildavsy, which describes arguments in the United States about nuclear power and air pollution in terms of competing ways of life: on one side egalitarian collectivists who use fear of disaster to argue against the kind of free enterprise that they also believe brings inequality; on the other side hierarchical individualists who want to defend free enterprise from public interference. Mary Douglas and Aaron Wildavsky, *Risk and Culture: An Essay on the Selection of Technological and Environmental Dangers* (Berkeley: University of California Press, 1983). Douglas argued that the various psychological ways of viewing risks that we look at elsewhere in this book are just tools people use to assert their cultural outlook, the means to an end. She was more interested in the ends. See Chapter 9, "Drugs," for Mary Douglas's ideas about risk as social control.

    See also *The Righteous Mind: Why Good People Are Divided by Politics and Religion* (New York: Pantheon Books, 2012), by Jonathan Haidt, who argues that our views about subjects such as climate change—along with most other political issues—come from a small set of tribal, moral outlooks. This notion has much in common with Dan Kahan's use of cultural theory. "Morality binds and blinds," says Haidt, meaning that morality, rather than scientific evidence, is what keeps your side together and what makes the other side seem stupid.

Communitarians, on the other hand, resent commerce and industry as "forms of noxious self-seeking productive of unjust disparity, and thus readily accept that such activities are dangerous and worthy of regulation," says Kahan.

None of this is unfailingly accurate, of course; there are plenty of exceptions. We are not arguing that climate-change skeptics have a monopoly on rudeness. But part of Kelvin's reason for rejecting climate change is that he doesn't like the political color of the solutions, which tend to mean more government, more regulation, more criticism of private enterprise. And if that means more control, it can't be true. Those on the other side, Kahan would say, believe what they do about climate change not necessarily because they're more scientific, but also because this gives them a chance to regulate private enterprise and control other people's behavior.

Big risks—those that might endanger hundreds, or thousands, or millions—are prone to cultural cognition partly because the evidence is bound to be uncertain to some extent, as there's no way of getting more data by rerunning experiments on the whole planet with and without an industrial revolution.

And partly because the crunch may be years away, affecting our children more than us—and people will have different views about how much the future matters. So climate change, like every other risk, is affected by the probability/consequence distinction, where some will worry more and others less about the potential costs to future generations (if, for example, they think those generations will be massively richer from economic growth and better able to deal with the problem). In fact, we can express our feelings about the future in a simple number, what's known as the "discount rate." If something that will happen in 50 years' time is just as important to us now as something that happens now, then our discount rate is 0 percent. We don't discount the future.

But this doesn't describe most of us. If you're offered $5 now or next month, most people think the $5 in the future is worth less—because it's in the future. They discount it. When deciding on health policies, the UK National Institute of Health and Clinical Excellence

(NICE) uses a default discount rate of 3.5 percent, which means that a year of life in 20 years' time is worth only around half of a year of life now. Hospitals deciding whether to use their limited cash to treat an older person or a younger one could say that the young person has a bigger future, and this counts for more. But if that future is heavily discounted, this works in favor of older people, because doctors are now required to put a higher value on their current year. Not having much of a future simply by virtue of being old is then less of a disadvantage in the competition for resources. With a 0 percent discount rate—which implies future years are as valuable as current years (and the future is not discounted)—more British National Health Service money and public health-care cash in the United States would be spent on younger people, who have more years left in which to benefit.

When economists look at climate change, they usually use a lower discount rate, such as 0.5 percent—to reflect people's concerns for the long-term future. Similarly, when deciding where to dump nuclear waste, if we had a high discount rate, we might just shove it in a bucket in a cave where it would probably be okay during our lifetime. But which discount rate is the right one can't be settled by science or statistics.

The UK National Risk Register gets around this by only being concerned with horrible things that might happen over the next 5 years.[4] In this project, people sit around mulling on possible causes of death and disaster and cheerily try to assess how bad they might be and how likely they are, and put them into a table like Figure 15. The scale of harm goes from 1 to 5 on the vertical axis. This is a human judgment, not a statistical fact, and so is the likelihood that many of these events will come to pass. In fact, the best that can be said in defense of these numbers is that they are better than nothing for governments that have no choice but to set priorities and make policies. They are intended to be fair working assumptions, not predictions.

These big hazards are hard to quantify with anything like the accuracy of the risk of heart disease, but they are risks that alarm governments. And again, it's clear that how serious you think they

How serious	1 in 20,000 to 1 in 2,000	1 in 2,000 to 1 in 200	1 in 200 to 1 in 20	1 in 20 to 1 in 2	Above 1 in 2
5				Pandemic influenza	
4			Coastal flooding Effusive volcanic eruption		
3	Major industrial accidents	Major transport accidents	Other infectious diseases Inland flooding	Severe space weather Low temperature and heavy snow Heatwaves	
2			Zoonotic animal diseases Drought	Explosive volcanic eruption Storms and gales Public disorder	
1			Non-zoonotic animal diseases	Industrial action	
			How likely		

FIGURE 15. The UK National Risk Register.

are may depend on your cultural outlook. Are you horrified by the threat of industrial action—militants bringing the country to its knees—or do you see no risk there at all, just ordinary people defending their living, a threat to no one if they're treated fairly? This is a repeated and important point in this book and in studies of risk generally: that some arguments about risk are not really about risk.

The UK National Risk Register only puts these threats into broad categories. To say, as it does, that the risk of an "effusive volcanic eruption" is between 1 in 200 and 1 in 20 means anything between a 0.5 percent and a 5 percent chance over the next 5 years. But there isn't much to base more precise figures on: the volcano in question, which is in Iceland, has gone off only twice in the past 1,000 years, so if we say that very roughly it goes off on average every 500 years, that's a 1 percent chance it will go off in the next 5 years. Mind you, it ranks highly on "impact": the last eruption, in 1783, killed 20 percent of the population of Iceland, wiped out their agriculture, sent clouds of sulphur dioxide across Europe, which fell as sulphuric acid and caused years of bad harvests, and so contributed to the French Revolution in 1789. One to watch out for.

When one of these crises occurs, or comes close, somebody has to be wheeled out to tell the public about it. There are manuals to teach you how to do this, full of commandments such as: listen to people's concerns, build on trust, express empathy, act fast, repeat messages, tell people what to do, acknowledge uncertainty, don't just reassure, commit to learning.[5]

Not much of this advice seems to have been followed when there was an outbreak of severe food poisoning in northern Germany in early May 2011, including cases of lethal hemolytic-uremic syndrome (HUS).[6] Scientists at a local laboratory tested some organic Spanish produce, found E. coli, and on May 26 announced they had found the source. Result: a widespread boycott of Spanish vegetables, despite film of the Spanish minister of agriculture desperately munching on organic cucumbers to show how safe they were. The problem was that, although the laboratory genuinely identified some E. coli, it was not the type that was causing the casualties, of whom 50 died. It was only once the Spanish industry had been well and truly messed up that the source was finally tracked down, in late June, to a shipment of Egyptian fenugreek seeds. And who remembers that?

An even more chilling case study in how not to communicate risk was the earthquake in L'Aquila, Italy, in 2009. After a series of small shocks and a local amateur with homemade equipment had predicted a major earthquake, there was a crucial meeting of experts on March 31, 2009, intended by the Civil Protection Agency to reassure the public. The meeting concluded that there was "no reason to say that a sequence of small magnitude events can be considered a sure precursor of a strong event," but at a press conference afterward, an official, Bernardo De Bernardinis, apparently translated this into the reassuring statements that there was "no danger" and that the scientific community assured him it was a "favourable situation." Go home, have a glass of wine, he reportedly said.

The succeeding events have the air of a scripted tragedy. At 11 p.m. on April 5, 2009, there was a strong shock, and families had to decide whether to stay indoors or spend the night out in the town squares—the traditional response to tremors. Families who heeded the apparently "scientific" reassurances remained indoors, and 309 people were subsequently killed in their beds when the devastating earthquake struck at 3:30 a.m., flattening many modern apartment buildings.

Six top Italian scientists and De Bernardinis were accused of manslaughter and stood trial in Italy. The scientists were not accused of failing to predict the earthquake, contrary to the impression given in some news reports in the United Kingdom, as it is acknowledged that this is currently impossible. The trial instead focused on what was communicated to the public. Did the defendants appear, or did someone appear on their behalf, to predict that there would not be an earthquake, despite their own knowledge of the uncertainties? If so, they had not read the manual.

A key question is what scientists and officials think people want to hear. One common refrain—often from scientists—is that people crave certainty and that that is unreasonable. The prosecution's claim in L'Aquila seems to have been, on the contrary, that the one thing they were sure they didn't want was to have uncertainty suppressed. The prosecution has been caricatured in some quarters as typical of a country that tortured Galileo, typical of a public demand for fortune-telling from necessarily uncertain scientists. The real issue is arguably almost the reverse.

One witness, Guido Fioravanti, described how he had called his mother at about 11 o'clock on the night of the earthquake, after the first tremor.

"I remember the fear in her voice," he said. "On other occasions they would have fled but that night, with my father, they repeated to themselves what the risk commission had said. And they stayed." His father was killed in the earthquake.

Another witness said: "[The messages from the commission meeting] may have in some way deprived us of the fear of earthquakes. The science, on this occasion, was dramatically superficial, and it betrayed the culture of prudence and good sense that our parents taught us on the basis of experience and of the wisdom of the previous generations." Otherwise, he said, they'd have slept outside.* As it was, they stayed in.

All the accused were convicted by a lower court of multiple manslaughter and sentenced to terms of six years in prison, but are, at the time of writing, appealing. One of them pointed out that he had previously identified L'Aquila as the biggest earthquake risk in Italy. The quality of the buildings also had a lot to do with the death toll.

Who else might be culpable for issuing reassuring statements? The British TV weather forecaster Michael Fish jovially discounted the possibility of a hurricane in October 1987, and the subsequent storm killed 18 people. In 1990, UK agriculture minister John Gummer was similarly reassuring about the safety of British beef when he force-fed his four-year-old daughter Cordelia a burger made of prime British beef in front of the cameras. Over 100 people have subsequently died of variant Creutzfeldt-Jakob Disease (vCJD) in the United Kingdom—although we think Cordelia is okay.

Of course, there's a difficult balance to be struck between reassurance and precaution. For every unheeded warning about a subprime crisis or cod depletion, there's an exaggerated claim about the potential dangers of saccharin or Y2K.

So although the Italian prosecution seems harsh, it's a salutary warning that people's emotions and intelligence should be treated with respect. People need full information and guidance for action, rather than just reassurance, and their concerns must be taken seriously.

---

* For a nuanced account of what went wrong in L'Aquila and what the scientists were actually accused of, see the article by Stephen Hall, "Scientists on Trial: At Fault?" in *Nature* 477, no. 7364 (2011): 264–269, www.nature.com/news/2011/110914/full/477264a.html.

11

GIVING BIRTH

"Uhh!" she said.

Norm shuffled his notes . . .

"Uhhhhhhghnnng!!"

. . . bit his lip.

"When will this thing be out?!"

"Erm, hold on," he said, shuffling. "Right, got it: Among 2,242 women with spontaneous onset of labor, the median duration of labor for those delivered vaginally was 8.25 hours in para 0, er . . . 5.5 hours in para 1, and 4.75 hours in para 2+ mothers." You're a para 0."*

He smiled.

"Oh God!"

Norm had spent days, days compiling data.

"Another one . . . "

He had footnotes on all his sources.

"Oh Goooodddddd!"

Everything written down alphabetically for ease of reference.

"Phu, phu, phu," she breathed.

He had been able to tell her at critical moments how likely she was to die.

She grabbed his sleeve.

---

* Meaning she's not had children before.

And had relative risk ratios *at his fingertips* for various anesthetics.

"Tell me one more . . . and . . . nggghhaaa . . . "

Threat of violence. . . . Here it was. Page 12. "Personal injury—low. Violent verbals—high (possibly of personal nature referencing anatomy)."

"You . . . dick! You stupid, stupid f . . . nnnnnngggghhh."

Excellent, he thought, spot on with that one, but then dropped his notes. It was about then she mooed . . .

"Mmmmmnnnnnoooooooo! . . . "*

Just as he was on the floor, scraping together the pages.

"Animal noise, animal, . . . " he mumbled, but the pages were all over the place.

"Nearly there," said the midwife. "You're doing brilliant . . . "

Well, thought Norm, depends what metric she uses. Performance, P . . . P. . . . He should have used staples. Time-wise, his wife was about average, truth be told. He was about to ask from under the instrument cart what the midwife meant when a note caught his eye on a meta-analysis of randomized controlled trials of vitamin K injections for newborns, he heard a gurgle, and the midwife tapped him on the back.

"Norm . . . Norm . . . It's a boy."

And there he was.

"A beautiful baby boy," she said.

And Mrs. N with a huge sappy smile, the baby in her arms, the midwife grinning like a slice of melon.

---

* Accounts of childbirth in literature seem to have been mostly sanitized affairs until the twentieth century, and mostly from a male point of view, like these from Kelvin and poor, befuddled Norm. Queen Victoria was unusually and famously forthright: "I think much more of our being like a cow or a dog at such moments; when our poor nature becomes so very animal and unecstatic." Quoted by Helen Rappaport in *Queen Victoria: A Biographical Companion* (Santa Barbara, CA: ABC-CLIO, 2003). But at some point women writers made the subject their own and the writing became more explicit: "The pain was no longer defined and separate from her but total, grasping, heating, bursting the whole of her, head, chest, wrought and pounded belly, so that animal sounds broke from it, grunts, incoherent grinding clamour, panting sighs." A. S. Byatt, *Still Life*, (New York: Scribner's, 1997). Or see Sylvia Plath, "Metaphors" (1959): "I'm a means, a stage, a cow in calf."

But now, "beautiful" . . . A tomato thrown at a wall, more like. Always tricky these aesthetic metrics.

Email from Prudence to Norm.
Attachments: *Health and Safety Executive: New and Expectant Mothers at Work: A Guide for Employers.*

Norm. Fantastic news. A precious time. Do be vigilant. Wise words in the attached Health and Safety Executive guide for new mums. See especially . . .

- Lifting/carrying of heavy loads
- Standing or sitting for long lengths of time
- Exposure to infectious diseases
- Work-related stress
- Workstations and posture

Love to all. Take care. Px.

"The thing with the missus's caeso," said Kelvin in the pub later, "was they couldn't put it all back in, did I tell you? No, not the baby, you jerk, the other stuff, hours, at it for hours they were, 'cause the caeso's the safe one you know, high-tech beep beep beep. Believe that and you'll believe anything. Anyway, whaddya mean, I said to the guy in the blue romper suit. You're not meant to have bits left over, how do you have bits left over? It's not an IKEA cupboard, is it, know what I mean? I mean, for starters, there's more room afterward, *comprendez vouz*? How can it not fit? Bloody shopping bag of emptied capacity, still no room and still fat and wobblin' after, how does that add up on volume, eh? But two hours, mate, botched job I reckon. 'Nother cigar? Surprised she didn't leak Guinness, not even as if yer even surplus caeso bits are useful or anything not like yer IKEA thingies them . . . What do you call them thingies? . . . You can't put caeso bits in a jar in the garage . . . Well, you could, but . . . now, hey, that is what they do, isn't it, hospitals, cut you to bits and put half of them in a jar, specimens, pickled . . . that's the scam.* Which bits do you leave out anyhow, biologically speaking, you considered that? Botched job, I reckon, what are the chances Norm? 'Nother pint? Go on you Zulu warrior, skull it!"

. . . .

MOTHERHOOD IS the most natural thing in the world. Is it also one of the most dangerous? In 2010 around 287,000 women worldwide died giving birth, 1 in every 480, equal to 2,100 MicroMorts.[1] That's similar to the average dose of acute fatal risk incurred by a British or American citizen about every five or six years, condensed, usually, into a matter of hours, as Norm says.

Even so, how useful are those mortality figures? How useful is Norm in a delivery room? That is, how often is the decision to have children—for those who have a choice—a calculation of risk, even though it might be the riskiest thing a woman does, especially in developing countries?

Danger that makes little difference to our decisions is a good example of probability's real-life limitations. If you think Norm sounds absurd or irrelevant here, it is for a reason. If emotion or compulsion seems more important than data as Norm scrapes around under the instrument cart, remember that about risk in general. "Who cares about odds?" some might say.

But only up to a point. The numbers tell stories that are extraordinary and appalling. At the appalling end, there is that global average. Worse still is that in some countries—Chad, Somalia—the risk is five times higher. In such places the maternal death rate in childbirth is a frightening 1 percent, and up to 2 percent in South Sudan, according to the Central Intelligence Agency. This risk of death of 1 in 100 births, or 10,000 MMs—might be considered the natural

---

* There was a scandal in the United Kingdom in 1999 when hospitals were discovered to be keeping body parts from surgery without patients' or parents' consent. For more information, see, for example, "Alder Hey Organs Scandal," http://en.wikipedia.org/wiki/Alder_Hey_organs_scandal.

rate, in that it probably applied in all of its brutality for millennia. Add up the risks in sub-Saharan Africa over a woman's lifetime, the World Health Organization says, and 1 woman in 39 will die during or soon after pregnancy. In the developed world the risk is 1 in 3,800. These developing-world risks are some of the highest in this book, although they are eclipsed by some of the unnatural, historical risks somehow contrived by medical authorities. The *natural* rate of death in childbirth is by no means the worst.[2]

To reduce mortality by three-quarters between 1990 and 2015 is a Millennium Development Goal. That goal will be missed. But the extraordinary part of that story in numbers is that it *has* fallen by about half, and at a pace often faster than the rich world ever achieved. Even the developing world has become a different world, if not yet different enough.

In the developed world, too, although maternal mortality has improved dramatically, the improvement has been slow, erratic, and marked by terrible suffering, often inflicted, as we now understand, by the medical profession. As any number of church memorials testifies, even among the upper classes childbirth was dangerous, perhaps even more dangerous than for the poor, who had the benefit of not being able to afford a doctor. About 150 years ago, 1 in 200 women still died this way in the United Kingdom, many from infection or "puerperal fever."[3]

In charitable laying-in institutions the risk was higher still, worse even than the Stone Age rate of 1 in 100. Giving birth in this period was safer at home, a fact recognized in 1841. Midwives were rightly feared, but were not nearly so deadly as doctors in some charitable hospitals. In Queen Charlotte's, London, "which possesses a great reputation as a school of obstetric practice," an extraordinary 4 in every 100 mothers died. Giving birth in Queen Charlotte's was a risk equivalent to 40,000 MMs, four times as bad as the natural rate of maternal mortality and far more dangerous even than a full year on active service in the British Army in Afghanistan (around 5,000 MMs in 2011).[4] A hospital could be more treacherous for women giving birth than a modern war zone. And the enemy? Doctors and hospitals killed most of these mothers, not labor itself. If we compare their mortality rate to the natural rate, it implies that three out of four maternal deaths at these institutions were directly caused by medical professionals more deadly than the Taliban.

In Vienna in 1848, a Hungarian doctor, Ignaz Semmelweis, was an exception. He compared two clinics, one for training medical students and one run by midwives. The medical students had two or three times the midwives' rate of maternal mortality, averaging about 10 percent, ten times the Stone Age rate.

In one month, December 1842, Semmelweis reported 75 deaths from 239 births, for a truly incredible rate of almost one in three mothers dying during or after childbirth. The hospital was a slaughterhouse.

Eventually, he discovered that the students—and their professors—went from handling corpses in autopsies to examining women in labor without thinking to wash their hands. He concluded that they were carrying "cadaverous particles" and said it would be safer to give birth in the streets. Some women preferred to do just that if they went into labor on the day of a student clinic. Semmelweis instituted handwashing with chlorine. The death rate dropped in one month from 18 percent to 2 percent.

His genius proved unhealthy. He was dismissed, moved to Pest, ever more disturbed by the general dismissal of his opinions, and wrote offensive letters to major obstetricians throughout Europe calling them murderers, with some justification. His behavior became more erratic and embarrassing, and in 1865 he was lured to an asylum, where he died two weeks later—by cruel coincidence from an infection after a beating from his guards. He was forty-seven. It took another thirty years and thousands of unnecessary deaths for the germ theory to become established and Semmelweis to be vindicated.

Data for maternal mortality in the United States before about 1915 are scarce. But afterward, there are stark differences. Irvine Loudon, an authority on the history of maternal mortality, compiled data to suggest that rates in the United States in the early twentieth century were far higher than in Western Europe, but with striking

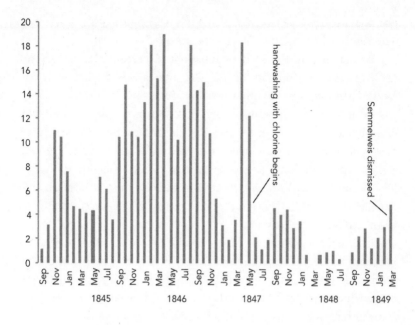

FIGURE 16. Historical mortality rates of puerperal fever at Vienna General Hospital maternity clinic, September 1844 to March 1849.

exceptions: "Midwives in the Kentucky Frontier Nursing Service, founded by Mary Breckinridge, traveled on horseback to assist with deliveries, which were all at home in a poor rural farming community with low living standards. Despite the poverty, maternal mortality rates were ≈ 10 times lower than those in the nearby city of Lexington and the United States as a whole."[5]

So in America, too, poverty could be a blessing—a rare exception to the general rule that better health is associated with higher social class. It wasn't until the 1940s that rates in the United States and Western Europe began to converge (although big differences remain).

And today?

First, you need to decide how to count a death as maternal. Some say it can be related to childbirth for up to a year after the birth, others say the limit should be 40 days. You also have to decide whether to include cases in which the childbirth exacerbates an existing condition that leads to death. For the United States, the United

Nations uses an expansive definition to give a figure of 21 deaths per 100,000 live births. The CIA says the same.[6]

But either way, the US rates are about 75 percent higher than the UK rates (12 per 100,000, on the UN figures) and higher than the rates of almost all of the other developed nations.

Even though the United States has seen a perplexing and not entirely explained rise in the past decade or two, by historical standards these rates in developed countries are extraordinarily good. Yet they are still equal to about 120 MMs (UK) (roughly the same for UK mothers as a motorcycle ride from London to Edinburgh and back) and 210 MMs (US) (around 800 miles on a US motorcycle, say from Nashville to Houston).

In the United Kingdom, there is also, still, a strong social class gradient, with five times the risk in lower economic groups compared with professional classes. Older mothers are also at higher risk. In the United States, it's race that stands out, with black American mothers about three times more at risk than white American mothers.

Taking the official UN data, in Sweden the risk is only 40 MMs, a third of that in the United Kingdom. As Semmelweis showed, cleanliness matters, and if you accompanied your wife as she gave birth in the United States you would probably have to wear a gown, whereas in the United Kingdom you can wander in from the garden. Even so, the official maternal mortality rate in the United States of 210 MMs—more than five times that of Sweden—puts it level with Iran in the international league table.[7] Although, inevitably, these figures are disputed.

The American comedian Joan Rivers said that for her the ideal birth would be if they "knock me out with the first pain, and wake me up when the hairdresser arrives," which is essentially what happened in Germany at the start of the twentieth century with a form of anesthesia known as "twilight sleep." Women could not even remember the birth. Anesthesia had been popularized by Queen Victoria in the 1850s—before which time women had no choice but to suffer Eve's punishment, as described in Genesis: "In sorrow thou shalt bring forth children."

The fear of giving birth is called "tokophobia." In the United Kingdom, there is an organization called the Birth Trauma Association, which says that anxiety is widespread and phobia not unusual, although in the extreme case of phobia this seems to be expressed in terms of disgust as well as concern about safety.

That they think it worth trying to overcome even this degree of fear shows how childbirth, in particular, presents a difficulty for any simple calculation of harm, namely: What's it worth? Because there is also, obviously, a benefit, a baby. The fact that millions of women and men accept the risks even if they have access to contraception and no qualms about using it (although see Chapter 8, on sex) shows that even severe risks of harm might be worth it.

In fact, there's evidence that people lower their estimation of the risk the more benefit they expect. This means not simply that for some people the benefit outweighs the risk and for others it doesn't, but that those who think the benefits are large also tend to think that the risk is *objectively* lower for themselves and everyone else. An imperfect rule of thumb might be: the more you expect, the less you worry. But why should the good reduce the bad? Compensate for,

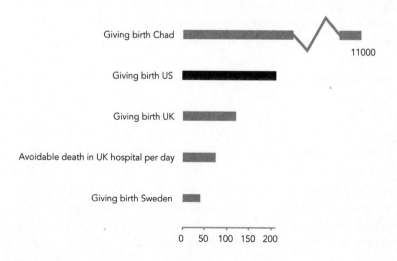

FIGURE 17. Average MicroMorts for mothers giving birth.

yes, but reduce? The psychologist of risk Paul Slovic calls this the "affect heuristic." If you like an idea, you find it harder to see how it might hurt you.

Would it be safer to press for a Caesarean section, as Kelvin's wife did? It is almost certainly a myth that Julius Caesar was born by this method, since it was then only used when the mother was dead or dying, and Caesar's mother seems to have been still alive when he was grown up. The first woman to survive a Caesarean was said to be a sixteenth-century wife of a man skilled in neutering sows, which supposedly gave him intimate anatomical knowledge. Regardless of the myth, Caesar's descendants are enthusiastic about the practice, since now nearly half of births in Rome are by Caesarean section, rising to 80 percent in private clinics. At 170 MicroMorts in the United Kingdom, the official record of deaths during Caesarean section appears to suggest that a C-section roughly increases the risk by a half.[8] But the risks are hotly contested, particularly as it is difficult to separate the risks of the operation itself from the risks associated with the reasons for having a Caesarean in the first place.

Of course, all sorts of things can go wrong that don't lead to the death of the mother but can nonetheless have a deep impact. The most common is postpartum depression, which affects around 10 to 15 percent of new mothers. It's another risk that runs back into the problem of how hard it can be to see past the benefits to the risks, especially when the benefits seem life-changing, or to take seriously the chance that our own feelings are a poor guide to how we will feel in the future. "I know that I want a baby, that I want it very much, that I feel happy at the thought of it, that the thought of not having a baby makes me unhappy, and that therefore having a baby will make me happy." Who is to tell you that you are wrong? But the risk of depression is not small. We can try to persuade people to take this risk more seriously, and perhaps we should, not least by offering proper help after the birth if problems arise. But we should not expect it to change people's hopes and behavior.

12

GAMBLING

Norm's gambling phase began after an unprecedented streak of roll-overs left the UK National Lottery Jackpot on £14 million, which was by strange coincidence precisely the sum that would buy every possible combination of numbers. So if he shelled out £14 million on tickets, winning would be a certainty.

Plus he'd also clean up all the lower prizes, which would yield a hefty profit. At which his eyes had an unusual—for Norm—Rambo look.

"Gosh," he said.

Though when he asked at breakfast how to raise £14 million, his wife didn't seem to hear.

"I said . . . "

"I know . . . "

"I've done the calculations," said Norm, "it's a rational strateg . . . "

"I think you may have missed the point of gambling," she said.

"It's a sure thing."

"As I said . . . "

Of course, there was the practical problem of how to buy 14 million tickets. At about 30 seconds per, that was still nearly 7,000 days at the machine, even full-time in Mr. Singh's. What finally put a spanner in Norm's plan was when he realized he had overlooked something.* Which is when his interest ended. No rational point, given the odds.

. . . .

How it all began, years previously, for the lad standing outside Ladbrokes betting shop, who was now desperately trying to sell to Kelvin even the Cocoa Puffs in his shopping bag to scrape together more money to gamble—at a slot machine at a family wedding reception:

"Oh, my lad! What a lad!"

"Play one for me will ya, sunshine?"

"He has the luck!"

"He's the knack!"

"Blond hair and blue eyes, it's angel luck!"

"Here, here's another, little fella, go on, son, go on."

And the gorgeous lights flashed, and the coins pumped and cheered.

How it ended:

"For 50p then? Cost a pound, mate, come on. Look: £1.23, on the box, come on, 50p!"

"Don't like Cocoa Puffs."

"What then? Milk? Fiver the lot? Come on. Please. A fiver. Lend us. I'll win it back. I'm due. Just a fiver, mate!"

And then, because the fare he'd kept safe in his back pocket hadn't been safe at all and eventually went the same way as the shopping, and there was the long walk home to find the note saying that she'd left after too many lies, and because it was a grand, more, blown in two days, a grand in her name they didn't have; and because he'd tried, God he'd tried, he'd even bought half the weekly shop to be safe and then sold it for f——all because the perfect moment to stop never came, never was a win big enough; because now his plastic is no good and he's crying again, and he sleeps in his clothes; and because his dad has already bailed him out but he blew

---

* Guess the flaw in Norm's plan. Then see the discussion in the second part of this chapter.

the bail-out and just the thought of that carves another hole in his soul, so back comes the sickness and self-loathing as he curses at the loser in the mirror, thinking everything he has is nothing and he is 24; because he is, in the end, lost; because of all this, but still somehow holding on to the wreckage of a daydream, he calls the Gamblers Anonymous free helpline sobbing in a piss-stinking phone box. Which was sad, because this was what he lived for.

But Kelvin didn't want the guy's Cocoa Puffs. He wanted Cheese Rind at 4–1 to get its snout in front on the big screen in the 7:30 Platinum Stakes at Reading, and as the money rode the track and the dogs strained, he was out of his skin with exhilaration. Norm said he was a fool, given the probabilities. Only, the next day, after Snow Queen topped the evening at an unlikely 7–1 over the 660-meter hurdles with fine late speed on the stretch, he won £81,281.52 on an accumulator. It was getting on for enough to buy a Maserati—if only he hadn't broken all his rules come the weekend and poured 40 grand down a women's tennis match on a Russian he'd never heard of because someone said Russians were good, but this was a donkey in a tutu. So? He'd won it first, which meant he lost nothing, and decided instead on a Porsche, and put it in a canal drunk that night having gambled on third-party insurance.

> Then said Jesus, Father, forgive them; for they know not what they do. And they parted his clothing, and cast lots.—Luke 23:34[1]

We can't know when people first started harnessing unpredictability as a fair means of making tricky decisions. But what went for Christ's clothing continues with some school admission lotteries, selection of juries, or deciding which survivor to eat in the lifeboat,* or sometimes buying a car.

---

* See, for example, Peter Stone, *The Luck of the Draw: The Role of Lotteries in Decision Making* (Oxford: Oxford University Press, 2011), which argues—with others—that the perceived problem with using lotteries for social decision-making is that they actively prevent reason from playing a part in the decision, but that this very quality has a "sanitizing effect."

Young US men sent to Vietnam were chosen by drawing from a box of capsules containing birthdays. Unfortunately in 1969 they put the capsules in the box in order of month but didn't mix them well, with the result that it was bad luck to be born later in the year: 26 out of 31 birthdays in December ended up drafted.[2]

The idea of a lottery should be to give everyone an equal chance. This randomness allows the intervention of whatever belief one has in God, fate, luck, or fortune, a process still seen in the bizarre rituals and mascots used in casinos and competitions. Norm, by trying to beat the lottery, is up to his old tricks again of trying not just to tame chance but to make it irrelevant. That would be power. But if you could do it all the time, would it be fun? Would it even be interesting? Maybe that's what Mrs. N is getting at. If you deny life its lottery, do you stifle life? Pitting ourselves against the gods is fun, but if you became one might everything become a little . . . predictable?

Devices that embody randomness have been used for leisure for at least 5,000 years. The ancient Egyptians, on a long winter's evening presumably, sat around and played board games in which moves were decided by throwing an "astragalus," a bone usually known as the "talus" in humans that sits just above the heel and under the two leg bones. As a game piece, it can fall on one of four sides. If you buy a leg of lamb, you can, with a little effort and mess, extract your own astragalus and use it for games or divination, just as the Greeks and Romans did. At some point, people started putting money on the outcomes of the throws, and the move from game-playing to gaming began. People used either astragali or dice: indeed, modern versions of the Bible now assert that "the soldiers gambled for his clothes by throwing dice."[3]

Gambling became so popular that the Romans tried to restrict it to Saturdays, but even the Emperor Claudius played obsessively and wrote a book: *How to Win at Dice*. People continued to make bets and quote odds: in Paris in 1588, you could get 5 to 1 against the Spanish Armada sailing to invade England, although this was probably a ruse by the Spanish. But what is remarkable is that in all this time, right up to the Renaissance in the 1500s, nobody analyzed gambles mathematically. Maybe they thought the outcomes were decided by some external force of fate, but it's now thought that the gap between theory and practice was too great, and the idea of putting a number on chance—the probability—just did not occur.[4]

The first book that started to work out odds was written in 1525 by an Italian, Girolamo Cardano, another obsessive gambler with no time for superstition.[5] He came up with the idea of counting the favorable outcomes—say, the six ways of throwing a 7 with two dice—and dividing it by the total of thirty-six possible outcomes, to find a chance of one in six of throwing a 7 with two dice. This seems obvious now, but it was a remarkable achievement. Although he got a lot wrong, too. He seemed to think that, just because an astragalus can land on one of four sides, each side is equally likely; but a brief experiment shows this is not the case (try it—astragali are uneven). He also thought that three throws of a die would be sufficient to give a 50:50 chance of a particular face, say a 6, a mistake that could have lost him a lot of money. You can try this as well. (The odds would be 50:50 only if there were no repeats.)

The Chevalier de Méré, a more perceptive gambler, in Paris in the 1650s, reckoned from his gaming that if he bet that he could throw a 6 in four throws, then the odds were slightly in his favor. Whereas if he threw two dice and bet he could land a double 6 in twenty-four throws, then the odds were slightly against him. By chance (or fate) he brought the problem to the attention of two of the smartest mathematicians at that or any other time, Blaise Pascal and Pierre de Fermat (he of the Last Theorem), who confirmed that he had a 52 percent chance of winning the first bet, and a 49 percent chance of winning the second,[6] so the Chevalier had actually got it right, from what must have been extensive and expensive experience. The Chevalier also presented the mathematicians with the "problem of points"—if a game has to stop before the end, in what proportions should the stake be divided up? A modern counterpart is the Duckworth-Lewis method for allocating runs when a cricket match is stopped, this time designed by statisticians—and so suitably incomprehensible.

The scientific assessment of odds then disappears for a few centuries, as the 1700s became a golden age of betting on the basis of gut feeling rather than calculation. It was the time of "eccentric wagers," such as a large sum won by the Count de Buckeburg in 1735 for riding a horse from London to York seated backward. There were also huge bets on cricket matches, and a predictable consequence was the match-fixing scandals that erupted in the early 1800s, with spectators at Lord's, the cricket ground in London, bewildered at the sight of two teams both desperately trying to lose.[7*]

As a result, bookmakers were banned from Lord's in 1817, and lotteries were banned in 1826. The Gaming Act of 1845 made gambling debts unenforceable by law in the United Kingdom. Cricket was turned into the archetypal gentlemen's game.

That is, until recently, when heavy gambling led to match-fixing, and, owing to the ability to bet on every detail of the match, players were bribed to mess up specific balls—an example of spot-fixing.[8]

Victorian morality abhorred gambling, and only with the liberalization of the 1960s did it begin to become a respectable pastime. Now, according to Gamble Aware, an organization in the United Kingdom that aims to promote responsible gambling and help prevent gambling addiction, 73 percent of adults in Britain gamble each year.[9] Even excluding the National Lottery, the figure is still almost half the adult population. The Gambling Commission estimates that they lost around £6 billion (US$10 billion) in 2009/2010—not counting the lottery. That's an average of £100 (US$160) per man, woman, and child, or around £200 per gambler. And that's the average. Most people don't lose anything like that, so some must be pouring it away.

Two statisticians once wrote a dense textbook on probability theory that they called *How to Gamble If You Must*, which led to many

---

* A similar extraordinary spectacle occurred at the 2012 London Olympics, when four badminton teams were disqualified for trying to lose in order to avoid a bad placement in the next round. See Paul Kelso, "Badminton Pairs Expelled from London 2012 Olympics After 'Match-Fixing' Scandal," *Daily Telegraph*, www.telegraph .co.uk/sport/olympics/badminton/9443922/Badminton-pairs-expelled-from -London-2012-Olympics-after-match-fixing-scandal.html.

disappointed purchasers and a subsequent change in title to the more accurate *Inequalities for Stochastic Processes*.[10] But what would be a good gamble now, if you must? We are not talking about corporate gambling on the financial markets, or games of skill such as poker, or the sports-betting organizations employing PhD mathematicians and using sophisticated statistical models, and not really the gambler on horse races who spends hours studying the form. We mean gambles for the fairly naïve punter.

First, lotteries. Lotteries prefer not to think of themselves as gambling, and the UK National Lottery does not even come under the Gambling Commission, but nevertheless takes in nearly £600 million (US$1 billion) each year. The jackpot in the main draw is won by matching 6 numbers drawn from 49, and as there are around 14 million possible combinations, that's a 1 in 14 million chance of winning from a single ticket. That's also about the chance of a 50-year-old woman dying from any cause in 15 minutes, or about the chance of being knocked off your bicycle and killed during a one-mile journey, or DS's chance of having a heart attack or stroke in the next 7 minutes—a bit more than the time it takes to buy a ticket. But with around 30 million ticket sales for each Saturday draw, on average two people should win each time. And the same number will die within a few minutes of buying their ticket. But nobody blames the lottery—though they would if it had been a vaccine.

If nobody wins, the big prize is rolled over. If this carries on, the jackpot could grow so big as to be more than the cost of buying every combination possible—say, if the UK jackpot ever reached more than £14 million. It would then become an interesting business proposition—as Norm realizes—to buy all possible tickets, and although this strategy would be difficult to organize, and isn't really for the naïve punter, it would win all the subsidiary prizes as well.

This has happened. In 1992 the jackpot in the Irish lottery stood at £1.7 million in spite of it costing only £973,896 to buy all the tickets and so be sure of winning. A Dublin syndicate managed to buy 80 percent of the tickets despite the efforts of the lottery organizers to stop them. And although they won the jackpot, they had to share it with two other winners; they would have made a loss, but with all the lesser prizes managed to make a small profit. And this is the flaw in Norm's plan. Norm knows that winning is a certainty—but he does not know who else might win, too. It's still a lottery.

That's the problem with this idea: massive payouts attract vast ticket sales, increasing the chance of having to share the big money. Take the US Mega Millions $1 lottery: in March 2012, the pretax jackpot stood at $656 million, although there were only 176 million possible tickets. In the end there were three jackpot winners—it would have been a high-risk gamble to try and buy all the tickets.

Win or lose, lotteries present an obvious opportunity for money-laundering, and the World Lottery Association even produces a guide on how operators can guard against this.[11]

Lotteries, often charitable, are one form of gambling in the United States that are widely permitted. America's history of periodic prohibition of gambling in one form or another, one state or another, has always created opportunities for those who are prepared to tolerate or encourage it, at least in a city or two, such as New Jersey's Atlantic City or Native American casinos. The commercial gaming industry is said by its industry body, the American Gaming Association, to have gross gaming revenue (the amount wagered minus winnings) of about $37 billion.[12]

But if you want the best betting odds, then visit one of the 150 or so casinos in the United Kingdom. You will still probably lose, but a European roulette table has only one zero, when the bank takes all the bets (an American table has two zeros), giving the house a small edge of 2.7 percent, meaning the casino returns, on average, 97.3 percent of money staked on roulette. This compares to the 45 percent return of the UK National Lottery, which doesn't sound great but is better than other lotteries.

Next, sports betting. Americans might be surprised to discover that betting on sporting events is a substantial part of the industry in the United Kingdom. Soccer, cricket, whatever, even other countries' sporting fixtures, are all part of this phenomenon. You might have to go to Nevada to bet on sports events in the United

States, but you can also go to just about any bookmaker in London to bet on the Super Bowl. You can also bet on incidents that happen during the games, back a point spread, or try to predict the total points scored in the game or how the first points of the game will be scored. Betting on horse racing and other sports is still a mainstay of the 8,000 or so betting shops in the United Kingdom, which operate at around an 88 percent payout, although Fixed Odds Betting Terminals (FOBTs) playing roulette at a 97.3 percent payout, also found in betting shops, are now more popular than horse racing and increasingly accused of encouraging problem gambling. FOBTs have a high payout, but the playing is so rapid and compulsive (DS can vouch for this) that they are staggeringly profitable—UK betting shops are only allowed four per shop, so outlets are proliferating in High Streets just to house the FOBTs. But these don't have the drama of an accumulator bet, in which a small initial stake accumulates over a whole series of connected bets provided they are all successful, such as picking nineteen correct soccer results in a row and winning £585,000 for an initial 86p stake.[13]

Lately, the Internet means that anyone can have a go, and Internet gambling has taken off, even though it is usually illegal and unregulated in the United States. The American Gaming Association says an estimated 1,700 offshore sites accept online bets, with the annual market estimated at $4 billion to $6 billion. In the United Kingdom, too, as if there weren't enough opportunities already, gambling is increasingly becoming a (legal) private activity conducted at home on the Internet—in 2008, 5.6 percent (1 in 18) of the adult population played online (non-lottery) games in the United Kingdom.

These websites may provide a link, hidden at the bottom of the page, to Gamble Aware and Gamblers Anonymous, because problem gambling is estimated to afflict around 1 or 3 percent of adults, depending on who you listen to. The American Psychiatric Association's psychological bible, *The Diagnostic and Statistical Manual of Mental Disorders*, has recently reclassified problem gambling as a behavioral disorder little different from a desire for, say, addictive drugs. If it is to be medically labeled an addiction, what next? Obsessive shopping?

Though it can clearly mess you up. Charles O'Brien, chair of the Substance-Related Disorders Work Group of the association, has said that "gambling behavior activates the same reward system in the brain that drugs activate."[14] This is a reward system from which people learn bad behavior, and the DSM-5, the latest edition of the diagnostic manual, says that your problem is mild if you experience four or five of the following, moderate with six or seven, and severe with eight or nine, for twelve months or longer, provided your behavior is not better explained by a manic episode:

1. Needs to gamble with increasing amounts of money in order to achieve the desired excitement.
2. Is restless or irritable when attempting to cut down or stop gambling.
3. Has made repeated unsuccessful efforts to control, cut back, or stop gambling.
4. Is often preoccupied with gambling (e.g., having persistent thoughts of reliving past gambling experiences, handicapping or planning the next venture, thinking of ways to get money with which to gamble).
5. Often gambles when feeling distressed (e.g., helpless, guilty, anxious, depressed).
6. After losing money gambling, often returns another day to get even ("chasing" one's losses).
7. Lies to conceal the extent of involvement with gambling.
8. Has jeopardized or lost a significant relationship, job, or educational or career opportunity because of gambling.
9. Relies on others to provide money to relieve desperate financial situations caused by gambling.[15]

These terse statements cover a pile of misery. So why do people continue to gamble in games of pure chance, when they know the house wins on average? The problem is that, although we may rationally know it's all just chance, and nothing we can know or do can affect the odds, we still tend to think that a good result is "due," or that really we have some control over what happens, and that "near-misses" are as important as they (truly) are in games of skill such as soccer.

There are a variety of problem gambling services in the United States, with the Oregon Health Authority being particularly prominent in providing free treatment and counseling, perhaps reflecting the "success" of the Oregon Lottery, which has been accused of relying on problem gamblers. Indeed, the Oregon Health Authority states that "service to out of state residents is permissible if the presenting gambling problem is reported as primarily related to an Oregon lottery product." There is only one UK National Health Service clinic for problem gambling, appropriately situated in Soho in London.[16]

So one risk of gambling is that people continue all the way to oblivion because they feel they can't stop. Do people without specialist knowledge also gamble as a rational substitution for buying a pension? If they do, they are truly irrational. But if we are not doing it as an investment strategy, and we don't wreck our health and family, perhaps it's not such a bad thing. In that case, gambling can be, well, fun.

So let's return to our question of the optimum strategy if you are an inexperienced gambler and want to try it, and fancy a £100,000 Maserati, like Kelvin, but sadly you have only a pound, and no background knowledge to allow you to use any skill. Let's also assume you are a cool, rational customer who wants the best odds (admittedly this is an implausible combination of characteristics, which is why Norm might be missing the point of gambling). If you buy a single UK National Lottery ticket, and if your choice of six numbers matches five winning balls plus the bonus number (a seventh ball drawn), then this generally wins around £100,000 and has a probability of 1 in 2,330,636.

Or you could go for an accumulator on the horses or dogs, like Kelvin: pick a meeting with six races, and in each race choose a horse at medium odds of around 6 to 1. An accumulator, in which the winnings of each race are passed to the next horse, will give you $7\times7\times7\times7\times7\times7 = £117,000$ if they all win. Given a bookmaker's margin of, say, 12 percent each bet, the true odds may be around 1 in 230,000, ten times as good as the lottery.

If you can find a casino that will let you bet just £1, place it on your lucky number between 1 and 36. When it wins, either leave the £36 there or move it to another number. When that comes up, too, move the £1,296 you now have to another number, or leave it where it is—it doesn't make any difference to the odds, but somehow it seems that the chance increases when the money is moved. When that comes up, you will have £46,656, so move it all to Red, and when that comes up you will have £93,312, almost enough for your Maserati. The chances of this happening, on a European roulette wheel with one zero, are $1/37\times1/37\times1/37\times18/37$, or 1 in 104,120, twice as good as the horses.

So roulette easily gives you the best chance of that shiny Maserati for a quid. Perhaps it's better to start saving.

13

AVERAGE RISKS

Norm's life lacked . . . what? He was thirty-eight years old and didn't know. But he knew missing when he saw it. This depressed him, although in truth not much. He was more moderately pissed off than depressed proper. That almost depressed him, too, in a middling kind of way. Where was life's, erm . . . you know? He tipped back on his chair and considered the curtains.

Like any average kind of guy, Norm knew in his heart that he was better than most. Proving it, that was the problem. In an effort at least to stand out he had taken up groovy habits like wearing striped socks or, for edge—you need edge—swearing a bit. Taking pleasure where he liked, whatever anyone said, he slipped into his shopping basket—alongside a two-pint carton of semi-skimmed milk, pre-packed sliced ham, breakfast cereal, and a chicken korma (mild) ready-meal—a bar of milk chocolate.[1] But that's how Norm was these days: more of his own man, picking out his own personal style at the Gap store. But still he lacked . . . you know . . .

He turned over an old envelope and wrote "NORM" at the top, underlined, twice. On the left, he wrote "Earnings," drew a line down the middle and scribbled on the right: "£28,270." He stared at the number.

It was weirdly familiar. He looked it up. He was right. It was the UK average.[2] How funny, to earn exactly the average for a full-time male employee.

"Height," he wrote, and then "5'9"." He looked that up, too, on the Office for National Statistics website. Also about average. Not too surprising, that one, he supposed.

Weight: Just over 13 stone.* He stared at the number, too, thinking. Then likewise looked it up. Again, average.

Weekly working hours: 39. Which turned out with a little research . . . he shifted in his chair and chewed his pencil.

Age at marriage . . . number of cups of coffee drunk per day . . . this was getting more than a little weird.

As he scribbled one stat after another on the envelope and then chewed the pencil as he typed his searches into Google, they all gave the same, uncanny answer.

Age at which he had his first child. . . . He was afraid to find out. Commuting time, hours spent watching TV, shoe size, number of fillings? He hated to think, but felt the answer in his bones already. How could one man be able to tick so many boxes, all in the middle?

For years Norm had yearned for the big event, the moment that would mark him out as unique, the way people do. He might even have worked harder for it but for the effort. Instead, he raised himself in his estimation by fine observations of the inferiority of the world at large: the grammar of the talking heads on TV, the speed of other motorists—damn idiots who drove fast, doddering ones who didn't. And all the while, here he was, bang in the middle. The implications were terrifying. He'd have to read the Daily Mail. He did read the Daily Mail (his wife's).

Could he fight it? He was tempted. Yes, to rebel, to throw off what seemed a fate of stifling, mediocre mediocrity, hypernormality, and do something uniquely impulsive and, oh, anything really . . . like, like . . . getting drunk.

Norm tapped his pencil on his teeth and stared again at the curtains. Was he humiliated? He wasn't sure. It was a lot to take in, this

---

* Or 182 pounds.

strange new status as some sort of, erm . . . He tapped his teeth some more. Then, abruptly, he stopped.

He sat back. He smiled. He put his hands behind his head. His smile grew, like that of a man who knew. He radiated satisfaction.

"Oh yes," he said, not unlike John Major. "Oh yes!"[3]

. . . .

THIS IS NORM'S APOTHEOSIS, the shining pinnacle of his narrative arc and moment of self-discovery—how bog-standard average he is. Quite simply, it makes him unique. That sounds absurd. How can anyone be uniquely average?

Norm can, because the average is not normally an individual quality; it describes everyone blended together. So it might apply to no one in particular, or at least to no individual, except, uniquely, to Norm.

He was always an average kind of guy—that much we knew—but we had no idea what a paragon of the ordinary he was. Nor did he. Plumb in the middle of the middle, the model of mediocrity, he probably drives a Ford Fiesta and goes on vacation in Ibiza. Nothing wrong with that, of course, but is it a special cause for satisfaction? And yet Norm smiles.

Perhaps because being average is strangely useful if you worry about the future, especially if you look to numbers for help. For all risks on which we stick numbers and probabilities are, in fact, averages.

So when it is said that you have a 20 percent higher chance of developing colorectal or pancreatic cancer if you eat an extra sausage every day, it does not mean that you personally face this increased risk; it means the average person does. So the average risk—or simply risk, for all risks are averaged over some group or other—describes Norm's future more reliably than it describes anyone else's, all those others—like you—who in some particular way differ from the average. Are they talking about me? On the whole, no. In Norm's case, yes. It's a bit like the child's fantasy that the world is designed around me. For Norm, it's sort of true. What an ego trip for a man who can't even make up his mind about licorice.

Norm—in a neat paradox for a man so unexceptional—is the archetype for man. He is no one and everyman. He is in no way outstanding, and, precisely because of that, outstanding. He is both prototypical and a one-off (add more oxymorons at will . . .). They should put up his statue—outside Walmart.

Not that this idea is without problems. We'll come to those in a moment. Meanwhile, allow Norm his strange glory.

The idea of the average person has itself become ordinary, but it is a statistical invention only about 150 years old. It originated with a nineteenth-century Belgian statistician, Adolphe Quetelet, who believed that the essential characteristics of the average man, "l'homme moyen," could be discovered by gathering data about the whole population, putting it on a graph, and looking for underlying patterns, peaks, or regularities.

Quetelet wrote: "If an individual at any given epoch of society possessed all the qualities of an average man, he would represent all that is great, good and beautiful."*

Cue Norm. As Mr. Average, he has more hope than most of being in tune with life's regularities. Although, if he had entirely average characteristics, he would also have about one testicle and about one breast. That is what happens when you add together a whole population. You find that an equal share of all the parts does not necessarily add up to a coherent human being. But Quetelet—a brilliant statistician—was not, in principle, deterred. He seems to have believed that the average was more than an abstraction, and thought that many averages represented genuine physical or mental capacities awaiting discovery, including moral capacities.

But Norm's satisfaction is everyone else's problem. Ordinarily, as we say, no one is precisely average in the many respects that determine the risks they face. Perhaps they are a little heavier, a little

---

* Quetelet derived l'homme moyen from the mean values of measured variables, which generally followed a normal distribution.

richer, or poorer, slightly more tense, sleep a little worse, are taller, slower, more sedentary, more tempted by cake, picked up some odd genes from a deranged ancestor, and who knows what else that might make a difference to their future, might tip the balance for or against survival to make the odds better or worse than average.

And there are other problems with defining life's prospects by the average. Some averages are ridiculous. Probably no one has the average number of feet. In fact, not even Norm can be average in every respect without running into a few logical absurdities. For example, he cannot be the average age all his life.

He is also unlikely to be the average weight for a man and the average weight for a thirty-one-year-old unless by luck these turn out to be the same. Different categories have different averages, and we all occupy many categories. Some averages are mutually incompatible. In other words, Norm cannot really be the average man, but can only ever be the average for some subset of men, sometimes a small subset, and by choosing one subset he ceases to be average for others. Very often the true average cannot exist as a man, or woman, at all. This points to a difficulty about all risk. It often describes the dangers that apply to someone who isn't there.

All this would be a terrible thing to say to Norm, so let's not break it to him. In any event, even if the average is theoretically imperfect and a bad fit for the complexity of life, it might work well enough to give Norm a practical steer as he decides what to do. We will see.

What indisputably does exist is variation around the average. Most people are not average, we all deviate from the norm/Norm, and these random deviations from Quetelet's *homme moyen* contain more of the real grit of life. They also change every individual's expectation of risk.

One of the best examples of bucking the kind of average that seems bound to determine life is the story of the American palaeontologist Stephen Jay Gould, diagnosed with abdominal mesothelioma at the height of a brilliant career, at a time when he had two young children. He was told that it was incurable and that the median* survival time after discovery was eight months. We might say that the risk—the average risk—of abdominal mesothelioma is death in eight months.

But this is not everyone's fate. It is the point by which half of all people who have it have died. And for those who haven't died by then, it is not the midpoint either. Gould lived for another twenty years and eventually died from an unrelated cancer. Averages of any kind, as Gould wrote in an essay describing this brush with mortality, are not immutable entities but an abstraction, and the true reality is "in our actual world of shadings, variation and continua."[†]

There are shadings, variation, and continua for all risks as well as for all averages. On average, men are taller than women. But then there is the Formula 1 boss Bernie Ecclestone, at about 5'3", and his former wife Slavica, at 6'2". There are, similarly, plenty of people out there who are Bernie Ecclestones of risk.

In fact, the average is sometimes misleading, not just about the odd individual but for the majority. About two-thirds of people in the United Kingdom have less than the average income (average in the sense of the mean). If we arranged the world's population in line, according to wealth, the average (mean) person would be about three-quarters of the way along the line.

Even MicroMorts and MicroLives cannot escape the problem of being averages. The average 7-MicroMort risk from skydiving arises mainly from the deaths of obsessive skydivers who do increasingly risky things. That is, the deaths are nearly all among experienced jumpers. The novice charity-supporting tandem parachuter may be

---

* Means and medians: for height, line up everyone from tallest to smallest and the median is the one in the middle; the mean is more like an equal share of all the height (i.e., what you get if you add up all the heights and divide the answer by the number of people).

† Gould tells his own story in a short, remarkable essay: "The Median Is Not the Message," http://people.umass.edu/biep540w/pdf/Stephen%20Jay%20Gould.pdf. For a longer, easy discussion of averages, see *The Tiger That Isn't: Seeing Through a World of Numbers* (London: Profile Books, 2010), by Blastland and Dilnot, or, in more detail, Sam Savage, *The Flaw of Averages: Why We Underestimate Risk in the Face of Uncertainty* (Chichester, UK: John Wiley, 2009).

taking just as much risk by simply getting drunk and walking home (but they probably wouldn't get the sponsorship).

Quetelet, no fool, knew all about this. He was every bit as sensitive to variations around the average. As he sat laboring over those reams of nineteenth-century data, the huge variation in human experience would have been obvious. The Body Mass Index, or BMI, by which we judge whether people are over or underweight, as well as normal, is also known as the Quetelet index.

So Quetelet stood between two ideas: one was the great scope of human variability; the other was the peculiar manner in which this variety seemed to contain an essence. As we will see in Chapter 14, on chance, the essence, the average, can be scarily predictable, but only at the right scale. This is the scale of whole populations, boiled down and their essence extracted. The problem is simply that this is not the scale on which individuals in all their variability live. Except for Norm. For the rest of us, when it comes to risk, none of us can attain the same state of Brahman-like self-knowledge as our hero.

Norm is a quintessence, even if sometimes a logically absurd one. Not everyone can say that. In fact, no one can say that. But is this enough for even Norm to be able to navigate safely through a world of hazards, using average numbers? Or does it mean that risk is never really his risk, or anyone else's, and even he is a fool to think otherwise?

14

CHANCE

Tall, chiseled, and gorgeous under long, dark hair, Kelvin Kevlin's older brother Kevin shared the same love of chance. Except that for Kevin, professor of social cognition at the Sorbonne and eye-candy TV pundit, now visiting Oxford University to deliver the celebrated Ronald McDonald public lectures, chance smiled on him.

He had a certain reputation for show-off controversy, confirmed by his recent book *God/I*, and it was this that drew the crowds. At the first lecture, a mathematician had been restrained from "punching the F . . . reudian lights out" of a psychologist. At the second, a Nobel laureate theoretical physicist working on string theory spat from the gallery while someone tore the dustcover from the upright piano and banged out some Wagner.

At the third and last, the crowd bustled. The talk, entitled: "I Am Not a Piano Key," was billed as an assault on reason. Rumor had it the professor would urge supporters to smash their cars into the walls of the college to prove they were alive, even if it meant death. Kevin, who spoke with fierce urgency and tended to jump straight into his lectures, pushed a strand of hair behind the ear, where a gold stud twinkled, and stepped up to the lectern.

"Reason is an excellent thing," he said, scanning the faces. "There's no disputing that. But reason is nothing but reason and satisfies only the rational side of human nature. The whole human life must include all the impulses, the will and the passions. It is not about extracting square roots."

Was it madness, genius, or fraud that danced in those eyes? He leaned on the lectern and glared at a front row of gigawatts of cerebral power.

"What does reason know? Reason only knows what it has learned—and some things it will never learn—while human nature acts as a whole, with everything that is in it, consciously or unconsciously. Even if it goes wrong . . . it lives," and he thumped his chest.

"Some of you look at me with pity; you think that an enlightened man cannot consciously desire anything bad for him. But I can. I might deliberately want what is bad for me, what is stupid, very stupid—simply in order to have the right to desire what is stupid and screw the obligation to desire only what is sensible. For this very stupid thing, this caprice of ours, preserves for us what is most precious and most important—this is our personality, our individuality."

Some in the crowd cheered, or jeered, it was hard to tell. The professor barely paused. Up in the gallery above the podium, the shouting was louder. A fight? Kevin thrust on.

"Give me every earthly blessing, a sea of happiness with nothing but bubbles of bliss on the surface, and even then, out of sheer ingratitude, sheer spite, I say 'Risk it all!' simply to introduce a fatal fantastic element, simply in order to prove to ourselves that we are still people and not the keys of a piano, which the laws of physics threaten to control so completely that soon we will be able to desire nothing but by the calendar."

By now, his hands were a conductor's, flicking, sweeping the air, his blue eyes enraged, his hair skipping, his voice rising.

"And that is not all: for not even the laws of physics can stop the play of chance that gives us limitless possibility and may even break the piano, that gives us freedom even to choose unreason. And even if we really were nothing but piano keys, even if this were proved by science and mathematics, even then we would not be reasonable but would do something stupid out of simple ingratitude, contrive destruction and chaos, only to make our point and convince ourselves that we are human and not a piano key!"

The gallery was heaving. A small cluster of young men moved toward the railing, clearing the crowd and pushing or pulling something heavy.

"If you say that all this, too, can be calculated and tabulated—chaos and darkness and curses—then people would go mad on purpose just to be rid of reason and make the point! For the whole of life is nothing but proving to ourselves every minute that we are human and not a piano key!"

In the gallery they bent down and lifted one end of the heavy thing. Its dark rectangular side appeared, poised, over the railing directly above the professor's head. There were shouts, a panic in the front rows of the standing crowd, a scramble, as the upright piano jutted further over the railing, some awkwardness as it caught, was heaved again, tilted, and then, to gasps and screams, reached its tipping point.

The professor did not look up. He ignored the noise. He poured out his argument. Reason fought against reason, climbing over itself into a hymn of hysterical praise to human impulse as the slow turn of the piano hastened, its great weight released into a drop, and its dark mass fell.

For a moment after the crash there was silence, except for the hum of piano wire and a fluttering of the professor's written notes in a haze of dust over a mound of splintered wood and the bones of the dark thing's cast-iron frame. Then, with the diabolical aura of the all-time lucky bastard, Professor Kevin swept back his hair, placed a foot on the debris, leaned at the retreated, gawping audience—and screamed with laughter.

A day later he was accused by a group of Oxford academics of shameless plagiarism of Dostoevsky, an accusation he mocked on the grounds that if it was shameless it was hardly plagiarism, as plagiarism tries to hide its tracks, whereas his was so blatant it could only be a tribute.* Also accused by the police of incitement to criminal behavior, he argued on the basis of the definitive critical appraisal of Dostoevsky's Notes from the Underground—a rant against rational egotism—that the argument therein was in fact parody and no incitement was intended, or for that matter understood, since no one did in fact smash into the walls of the college. He was let off with a caution. He refused to assist the police inquiry into his attempted murder by a group of medical students, an accusation later reduced to a charge of criminal damage, although the suspicion persisted that the whole incident had been contrived. A day later he resigned, without reason, he said.

. . . .

NONE OF US KNOWS what will happen tomorrow, let alone years hence, and many don't want to. Like Professor Kevin. In a fine family tradition, he prefers life instinctive and messy. Choices are not calculable anyway, so why try?

Others—like Prudence—want control over their future and take every care to get it, if they can.

Key to both Kevin's hopes and Pru's fears is chance. He loves its power to throw up the unexpected. She hates it for the same reason. Chance is the rogue that threatens either to wreck their best-laid plans or, as if by magic, pull a rabbit from the hat. (Norm, standing between them, thinks he can play the odds—to bet whether rabbit or wreckage is more likely.)

But what is chance? Philosophers have wrestled with the problem for centuries, from declaring chance all-powerful to wondering if it exists.

---

* The Underground Man, from whom Kevin takes most of his speech, argues for the freedom to reject 2 + 2 = 4. "I admit that twice two makes four is an excellent thing, but if we are to give everything its due, twice two makes five is sometimes a very charming thing too." Fyodor Dostoyevsky, Notes from the Underground and The Double (Harmondsworth, UK: Penguin, 1972). Other Dostoevsky characters rebel in similar ways, in order to affirm their individuality.

We're going to put the question in an unusual way, with a dark edge of practicality, by asking: Why does the piano miss Kevin? It wasn't meant to. He was all but a dead man. By what means did he survive?

Kevin himself answers by linking chance with free will as two expressions of the glorious muddle in life that makes us humans and not machines. He sets this freedom against an idea of reason that he seems to equate with soulless determinism. For while strict necessity of cause and effect implies only one possible future, chance and free will in their different ways can turn away from this necessity. In Kevin's eyes, then, the piano misses him because all of life, material and human, must have a "fantastical element" that breaks the chain of strict causality. This is the way he thinks life ought to be— and the way it ought to be lived—even if he betrays a doubt or two about whether that's how it truly is.

The dictionary is not so dramatic, defining chance mainly as a prosaic possibility: What chance of rain? In everyday use we often go further and give chance an improbable twist, saying that to "take your chances," for example, is to be something like a gambler, a risk-taker who knows the odds are against success, but what the heck? If you wanted a story about chance, someone might tell an improbable tale about an unintended event on which the whole plot hinged, a strange coincidence or happenstance, the more fantastic the better.

And if chance brings calamity, when star-crossed Shakespearean lovers die for want of knowing that one of them is only feigning death, or for the jealousy set off by a dropped handkerchief, we call it tragedy, in which fate is fantastically cruel and chance is about the small but fatal detail.

In all these other, fantastical uses, chance is a synonym for outrageous luck or bad luck, a long shot or a fluke. But none goes as far as implying, like Kevin, that chance is capable of breaking the chain of causality. They only say that chance means a chain we didn't expect. So the piano misses Kevin because of accident, as it turns in the air past his gorgeous black hair, owing to the way it was tipped from the balcony as the frame snagged, just when death seemed a certainty. The cause was in some small detail of the angles, masses, and forces, and his own antics at the lectern.

One response to life's uncertainty is to argue that there are deep, invisible causes for everything, unlikely or not. In the words of St. Augustine: "We say that those causes that are said to be by chance are not non-existent but are hidden, and we attribute them to the will of the true God." The piano misses Kevin because God looks kindly on him. Though why, we'll never know.

The German poet Friedrich von Schiller said similarly: "There is no such thing as chance; and what seem to us merest accident springs from the deepest source of destiny." All things happen for a reason.

A similar reaction to chance is superstition, by which people hope that rituals and totems will turn fate to their advantage, perhaps by finding favor with the gods, perhaps through mystical connection with natural power, and where today's lucky sporting mascots are the sanitized versions of Aztec human sacrifice. Fate calls the shots, but likes a bribe. The piano misses because Kevin must have slaughtered a sheep at high tide at midnight.

Even 2,000 years ago there were "rationalists" who scorned these beliefs. Oddly, Kevin would have liked their style. In Roman times, three 6s from three dice was known as "the Venus throw." Cicero said: "Are we going to be so feeble-minded then as to aver that such a thing happened by the personal intervention of Venus rather than by pure luck?" He taunted astrologers with relish.

Cicero, like Kevin, defends freedom from necessity. In Cicero we see this as rationality at war with superstition. Kevin sees it as antirationality in favor of chaos and impulse.

As the scientific enlightenment developed in the late seventeenth century, earthly causes challenged mystical ones, and the extraordinary explanatory power of Newton's laws of motion led to the belief, in the physical world at least, that all matter moved like clockwork: if we knew precisely the position and motion of every atom, then in principle we could predict exactly what would happen. Thus again, the piano misses Kevin because the precise chain of causality between atoms, forces, and initial conditions, following given laws, plots its course, just shy of his nose. And again, there is nothing inherently random about it. Any uncertainty in this world is "a measure of our ignorance"—as the statistician Pierre Laplace said—ignorance not of God's will this time but of the state and laws of nature; not so much "can't know" as "don't know."

Two words summarize these two types of uncertainty. The first is "aleatory," used to describe the uncertainty before a coin is flipped when we simply cannot know what will happen, often known as chance or randomness. The second is "epistemic," used for the uncertainty when we have flipped the coin but not looked at it, in which case we don't know but in theory could—more commonly called lack of knowledge or ignorance. Although there are inevitable complications: What if it's a two-headed coin? Then what you thought was chance was tainted by ignorance.

The "don't know" that applied to human knowledge of the mind of God and the "don't know" that described our limited understanding of nature's clockwork implied that with more knowledge we would uncover the rigid determinism that explained human behavior fully. Either way, everything had a cause that began outside ourselves.

This spooked people. Both kinds of ignorance—of divine cause and physical cause—seemed contrary to our internal sense of free will, and neither our own life nor other people's behavior appears to us to obey the mechanics of a pendulum. So at this stage (and ever since), there were fierce arguments in both religion and science about whether people's sense of their own freedom of choice was an illusion.

Then came the grand age of statistics, when people were initially drawn further into the net of predictability. Beginning in the early nineteenth century there was an obsessive listing of deaths, crimes, and, particularly, suicides. And what emerged was an extraordinary regularity, even from the turmoil of individual decisions, an order that seemed to emerge as if by a natural law and so raise doubts about the scope for chance. Because despite the choices of millions, each in unique circumstances, there seemed to be some regularizing force that led to an almost constant number of suicides each year. Other patterns also emerged from the data. Francis Galton, a cousin of Charles Darwin, said famously: "Whenever a large sample of chaotic elements are taken in hand and marshalled in the order of their magnitude, an unsuspected and most beautiful form of regularity proves to have been latent all along." Some came to the conclusion that these predictable patterns meant that chance could not exist.

But although there was order in human behavior en masse, there was still mystery about how this order would shape any individual. Yes, there would be a roughly predictable number of suicides, but still it was not possible to say by whom. The average, the trend, and the distribution were indeed observable, but the particular was not predictable.

This led to the idea that we are all random deviations from Quetelet's average man or *homme moyen* (described in Chapter 13). On this reading of chance, were the lads to throw a piano off the balcony an infinite number of times while Kevin gave his lecture, then, on average, Kevin would be dead. He survives because he is a random variation from the average, and more than happy to be such a deviation.

There were strikingly similar developments in the physical sciences. Quetelet described his work as "social physics," and his ideas, published between the 1820s and the 1860s, may have influenced James Clerk Maxwell's development of Daniel Bernoulli's kinetic theory of gases, which describes the overall behavior of a gas as akin to a vast blizzard of colliding individual particles. This parallel between social and physical science doesn't receive the attention it deserves, but it is a remarkable intellectual coevolution that seems to offer a way of comparing human experience with the behavior of matter. Here, too, although it was impossible to predict the motion of each of the zillions of molecules bouncing around in a gas, even if this motion was in theory fully determined, probability could successfully explain the overall movement, just as it can describe overall patterns in people's behavior. In fact, there is an argument that without probability we would be unable to appreciate large-scale change in people or in physics.* We would sit there watching

---

* J. Robert Oppenheimer explained: "If we insisted on the detailed description of the motion of individual molecules, the notions of probability which turn out to be so essential for our understanding of the irreversible character of physical

a zillion random events among humans and molecules, with little sense of a bigger picture.

In theory, again, it was *possible* for all the molecules to wander off suddenly in one direction, as when the Improbability Drive in *The Hitchhiker's Guide to the Galaxy* was "originally devised by physicists to transpose the underwear of a hostess at a party several feet away from her body."[1] In practice the behavior of the whole gas was predictable even if the behavior of the parts were in practice not. So in both the social world of people and the physical world, the aim of predicting what would happen to each little piece, whether a human or a molecule, was given up, but a new order was found in the predictability of averages.

Perhaps the best image of the behavior of molecules is provided each Saturday evening, when 49 balls are banging around in a bucket, watched by millions of people in the United Kingdom clutching their National Lottery tickets—or 59 white balls in the US Powerball lottery, plus 35 red balls for the Powerball itself—and hoping their lives will be changed. Six balls are chosen for the jackpot—as we have seen in Chapter 12, it makes no difference which set you choose (1, 2, 3, 4, 5, 6, or any other combination), the chance is still around 1 in 14 million that your choice will come up, or about 1 in 175 million for the Powerball. Websites keep track of how often specific numbers come up, with 38 having topped the UK list for number of appearances for ages. Surely the law of averages says that 38 is due for a rest and we should keep away from it?

But this isn't how the law of averages works. What it means is that over time, even though each draw is completely unpredictable

---

events in nature would never enter. We should not have the great insight that we now do: namely, that the direction of change in the world is from the less probable to the more, from the more organised to the less, because all we would be talking about would be an incredible number of orbits and trajectories and collisions. It would be a great miracle to us that, out of equations of motion, which to every allowed motion permit a precisely opposite one, we could nevertheless emerge into a world in which there is a trend of change with time which is irreversible, unmistakable and familiar in all our physical experience." J. Robert Oppenheimer, *Science and the Common Understanding* (New York: Simon and Schuster, 1954).

and not influenced by what has happened before, the spread of the numbers of appearances of the balls has a regular pattern, as shown in Figure 18. All these unpredictable events, when put together, turn out to have their own structure and even their own shape—the beautiful normal curve observed by Quetelet when he measured people's heights.

So although 38 has popped up 241 times in the UK National Lottery—nearly half as many times again as poor old number 20's pathetic 171 appearances—this is exactly what could have been predicted back in 1994 using probability theory alone. Of course, we could not have predicted which numbers would be top and bottom, just that one number would have about 240 and one around 170 appearances.

Another wonderful illustration of the patterns of chance comes from a contender for the most boring book in the world—*A Million Random Digits*. The title says it all—page after page of numbers, with no detectable order, each one utterly unrelated to those that come before. It begins with "1" and finishes with "8," an ending as

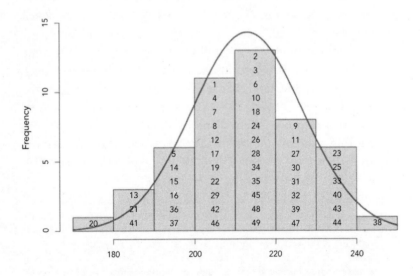

FIGURE 18. The number of appearances of balls 1 to 49 in 1,740 UK National Lottery draws, November 1994 to August 2012. The normal curve of the line is the predicted distribution based on probability theory.

unpredictable as a good Agatha Christie whodunit, and the authors noted that "because of the very nature of the tables, it did not seem necessary to proofread every page of the final manuscript." The audio version would do wonders for insomniacs. And we're still waiting for the German translation.[2*]

But there are gems hidden in all this randomness. We can expect the sequence 1, 2, 3, 4, 5 to occur ten times (and in fact it occurs eleven times). And we would expect events that have a one-in-a-million chance, such as the same digit occurring seven times in a row—to turn up once. And, satisfyingly, there is exactly one example of this—6, 6, 6, 6, 6, 6, 6. And we can even calculate the probability that there would be no such sequence, which turns out to be 37 percent, so we could have been unlucky.

Probability theory provides a practical way of dealing with the randomness in life—but even the patterns of pure unpredictability don't really tackle the fundamental question of whether true, absolute, irreducible chance really exists, or whether everything is at some level determined. If we knew everything about those bouncing lottery balls, could we predict which one would drop out? Sadly not, for two main reasons.

First, quantum mechanics came along in the twentieth century and said that chance did indeed exist, at least at a subatomic level. Essential parts of the subatomic world could only be described as a probability. Heisenberg's Uncertainty Principle says we can't know everything about a subatomic particle—in terms of both where it is and where it is going. Newton's laws of causation were inadequate.

Second (and in practice more important than the quantum uncertainty), there is also an effect that throws all expectations into the air. This is chaos theory, in which, even if a system were

---

* The numbers for the book were generated from an electronic pulse and hence incorporate genuine unpredictability. In contrast, "random" numbers produced by modern computers come from logical, but very complex, arithmetic operations, and hence follow a sequence that is in principle predictable and reproducible if the "code" is known—such numbers are known as "pseudorandom."

completely deterministic, in the sense that there was no randomness at any point in the chain of causation, some systems—such as the weather—might nevertheless be so dependent on, and so affected by, the tiniest difference in where they start, a difference undetectable by people, that we have no idea how they will turn out.

The standard image is of a butterfly flapping its wings and thus causing some faraway storm. A real example is Clint Dawson's experience of forest fires. He is an FBAN, a fire behavior analyst, and his job is to try to predict the course of forest fires in Colorado. In 2012 the fires began to behave with dangerous unpredictability, often burning bigger than before. His computer models of what a fire would do became less accurate. The reason? A series of tiny changes in initial conditions, among them that no one had thought—why would they?—to factor in a change in the behavior of beetles, of all things. A winter influx of pine beetles had desiccated the trees, making them more flammable.[3]

The chaos idea means that clockwork-like systems can be unpredictable simply because we can never even know exactly where they start, let alone exactly what they will do next. Causation is too complicated. Such systems might as well be ruled by chance after all (whatever that is). Had a butterfly landed on the piano just as the students moved it, and distracted one of them, it might have made only the tiniest difference, which to Kevin would have been a difference of life and death.

Do these physical limitations on what we can know, the unfathomable complexity of chaos, and quantum uncertainty, matter on a human scale? Is it odds-on that Kevin's trust in some fantastical element will be his doom one day? Or are these strange forces only of theoretical importance? How much faith would you have in uncertainty if a piano loomed over your head?

All risk is chance. And chance remains mysterious. Whether it originates outside people in the deep structures of matter or if it's all in our heads hardly matters to the uncertainty of living. The practical question is what to do about that: whether to be more like Prudence, the Kevlin brothers, or Norm.

TRANSPORTATION

Norm shuffled along the aisle until he found the ideal space, four empty seats, one at the window to watch the station do its strange retreat; just him, his thoughts, a book on the table. He liked trains—and their low MicroMorts.

A short, stout, grinning woman with bleached hair wearing an outsize white T-shirt bumped her way into the seat opposite and stuffed her backpack under the table between her knees. Norm pulled in his feet and reached for his book.

Then he noticed: a wide, bright, red-and-white-striped sweatband with a (badly) stitched-on smiley, halfway up her forearm.

Norm was a man of the mind. *Four-inch sweatband.* He went through life preoccupied, tolerant, rational, and relaxed . . . *red-and-white-striped* . . . though open-minded, naturally, about the potential for personal enrichment via others' experience . . . *stitched-on smiley* . . . and believed each to their own, live and let live, John Stuart Mill, etc. . . . *halfway up!*

His eye caught the emergency hammer. He sank in his seat at the thought of hours of "chat" with a four-inch, red-and-white sweatband and stitched-on smiley *halfway up* the kind of forearm that took its pleasures—he didn't doubt—tearing off testicles.

Thinking fast, he took a firm grip on his emotions and began to calculate the risks: (1) Assume successful avoidance of eye contact; (2) Combine low-to-moderate chat hazard, with (3) significant threat of, if not spontaneous psycho violence, then at least embarrassment-grade nutterness; plus (4) (he winced, but there it was) probable class and diction issues—Glaswegian?—and arrived at a 51 percent probability of hell. Approximately.

She reached into her backpack and pulled out . . . what?

Fifty-two percent, thought Norm.

A large red paper napkin that she unfolded onto the table . . .

Fifty-four percent.

. . . Spread well onto Norm's side . . .

Fifty-eight percent.

Reached down again, paused, and placed one-by-one on the napkin: an all-day-breakfast mixed-triple-pack sandwich of mayo, spicy sausage, and bacon and tomato . . .

Sixty-eight percent.

A small pork pie . . .

Seventy-five percent.

Crisps—or potato chips—and a Ripple, a Crunchie, and a Mint Aero.*

Eighty-two percent.

And a can of Strongbow hard cider.

Ninety-five percent.

Norm whimpered.

She rotated the Ripple 180 degrees.

Ninety-nine percent. Holy shit.

Norm was a dead man. Paralyzed. Waiting for the cobra to strike. Surveillance blown, he stared floodlights. The only option was to get off the train. Was ever a woman of more menace? Every move ended with a pause—and another smile. Every smile ratcheted Norm's unblinking terror.

She picked up the can, stared at the label, smiled, and pulled the ring.

It hissed.

She paused.

---

* All British candies.

She tore it back—and set the can carefully beside the Crunchie with her right hand—*sweatband halfway up*. She paused. *Stitched-on smiley*. She lifted the can and drank, deep and long. He watched her throat's gristly dance. She put down the can, paused, and smiled.

Please, he thought regarding the pork pie, let her not spit. Then a more awful thought crossed his mind: What if she offered . . .

She caught his gaze and tilted the can with a quizzical eyebrow. Oh God.

Please don't say *swig*, thought Norm. "Erm, thank you, thanks. I won't actually. Just had breakfast, big one, already, really, coffee, tea . . . orange juice, coffee, honestly. Already. Thanks."

She nodded, and turned to face the man sitting next to Norm, hidden, but for his fingertips, behind the *Financial Times*.

"If you don't mind me saying," she said in cut-glass English, "it's probably a weakness of demand, at least in part, don't you think? A dab of Keynesian stimulus wouldn't be out of order. The economy, young man. Well, you can tell by the data on corporate investment that it's as flat on its back as a tart."*

The *Financial Times* snapped straight.

She smiled and looked down at the tabletop, placed her palms either side of her selection, studied it, picked the Crunchie, unwrapped it with delicacy, raised it to one nostril, and inhaled. She moved the Crunchie toward the side of her mouth, closed her eyes, and then, as if amplified, pulverized the end between her molars.

She sat back, opened her eyes, and crunched, like a Scottish lord pacing the long gravel drive of his Highland estate, and smiled again into the middle distance.

. . . .

WHAT'S HAPPENED TO NORM? The rational paragon just became as meek as a mouse. There he was, nicely archetypal, suddenly turned paranoid because of a wristband, an accent, and a pork pie in his

---

* About which she was at least half-right, though whether Keynesian stimulus was the solution is not for us.

personal space. Has he lost the plot? Well, yes. But then, we do. People are as inconsistent about danger as other things—often for good reason.

They may, for example, lose their nerve. John Sergeant was a BBC correspondent who covered conflicts including Vietnam and Northern Ireland. Then he was taken hostage in Cyprus and held at gunpoint for thirty-three hours. After that, he said, he didn't want to do it anymore. He realized that he was "extremely frightened" and went into reporting politics.[1]

Something similar happened to David Shukman. He'd reported from war zones for fifteen years. One day, asked to board a worn-out-looking helicopter from Tajikistan to Afghanistan shortly after 9/11—he refused. And that was that. The more he reflected on his assignments, the more nervous he felt. He also thought more about the price paid by his family.[*]

Stories allow people freedom to change. Probabilities could be taken to suggest that people should stay the same as long as the numbers stay the same. And one response to people who change their minds is to say that it proves we can't judge danger sensibly. For it's not as if the outside world has changed, as if train travel for Norm is more dangerous now, or war reporting—or war—wasn't dangerous enough before. In terms of probabilities, the risks to Norm are the same before and after the pork pie. All the change is on the inside.

There are two simple justifications. The first is new information (see also Chapter 14, on chance, about epistemic uncertainty). Norm learned something about the risks of train travel that he hadn't known or imagined before. Doctors, governments, and health departments also all revise their advice about risk as they learn. But true risk—whatever that might be—is unknowable, since our understanding can always be improved. Always lay your sleeping baby

---

[*] One of the best-known literary losses of nerve also comes in an account of war. Yossarian in Joseph Heller's *Catch-22* sees an exceptionally gruesome death and decides that the enemy is anyone trying to kill him, and this includes his own side. The world becomes more deadly even as it remains the same.

face-down, they said with confidence, until the evidence changed 180 degrees. "If the facts change, I change my mind," said the economist John Maynard Keynes, and the facts about risk are seldom final.

The second justification for changing your mind is that you see danger in a new light, like David Shukman did when he thought more about the pain his family felt, or as Norm does when his prejudices are stirred.

So to expect lifelong consistency from Norm would be inconsistent: it's not normal. Norm's basic approach to life—that reason and calculation are the one true way—will survive, you'll be glad to know, but perhaps with more sense of his own ups, downs, ignorance, and weaknesses. All the same, he has taken a knock.

But he would still want facts about the quantifiable risks of traveling, and here they are, first by train. Roads and air follow.

TRAINS

The hazards began—badly—on September 15, 1830, with the opening of the Liverpool to Manchester railway. William Huskisson, a leading English politician, went to greet the Duke of Wellington at his railcar and was flattened by George Stephenson's *Rocket* coming up the other track. His leg was "horribly mangled," and he died a few hours later. Oddly, his death attracted huge and welcome publicity to the railway. This lesson—that it is safer to be inside a train looking out than outside looking in—is still valuable.

With qualifications: the inside of a train is also a public space. Of itself, the presence of other people—wristbands included—has only the tiniest effect on your chance of death or injury. Norm's personal space invasion makes him quake, but with no basis in mortality statistics—murders on trains are a good plot device but extremely rare, although the headline "Victim Describes Shock of Being Randomly Stabbed by Deranged Woman with Steak Knife on Subway" speaks for itself.[2]

For lesser violence, the risk is also real and also small. Amtrak does not publish statistics about violence on US trains, but 3,300 assaults were recorded on trains in the United Kingdom in 2010. Even so, this is 1 in 400,000 journeys, and so if you traveled every day on a UK train you might expect to be assaulted once around every 1,000 years, if the risk were equally shared.

Still, this is little comfort to Norm, who might fear that this is his 1,000-year reckoning. Or maybe softer factors than hard data explain his sudden feeling of peril. For example, perhaps prejudice distorts his sense of the odds. Or perhaps he's disconcerted by those "class and diction issues," as Norm puts it, just as other people feel threatened by awkwardness like bad language, noise, being pestered or leered at, or the smell of someone's burger. All these things happen, though none show up in the statistics for danger. So is it this—unpleasantness rather than physical injury—that people are really worried about when they say they feel unsafe on public transportation? Is background irritation part of what makes the company of strangers menacing, even when the risk of violence is remote? Norm is afraid because he can't predict wristband's behavior—"psycho violence" or "embarrassment-grade nutterness?"—and we feel more wary of strangers than friends or relatives on the backseat of the car (depending on who your friends and relatives are) for exactly this reason—we don't know quite what to expect. In other words, the risk arises from uncertainty linked to unfamiliarity in any individual case, not from statistical likelihood. As one research paper put it: "Feelings of anxiety and psychological factors act to make some people feel uncomfortable on public transport and . . . this acts to increase perceptions of poor personal safety."[3]

Other influential factors named in the study were gender—women tend to be more anxious than men—and "the actual experience of a personal safety incident." Mind you, even the most pronounced of these influences was found only to tinker at the edges of what makes the most difference of all in how we feel, and that is the accident record itself. So the numbers do matter.

So, more numbers. Despite William Huskisson's misfortune in 1830, train travel grew massively in the subsequent 180 years. In Great Britain in 2010 there were 1,400 million journeys, up from 800 million 30 years ago: that's 4 million journeys a day, totaling 54 billion kilometers (or about 33.5 billion miles) over the year.[4] This means that each one of 60 million people does on average 900 kilometers (560 miles) a year, or around 10 miles a week, although this is one of those wonderfully misleading "averages" that reflects the experience of nobody. Who travels 10 miles a week by train? There is huge variability, ranging from the dedicated commuter to someone in the countryside, cut off from trains by the declines of train travel in the 1960s, to whom a trip on the rails is a rare treat (or not).

At the bottom of this distribution will be sufferers from "siderodromophobia," the inordinate fear of trains, but most people feel a reassuring solidity and familiarity about train travel, as does Norm, usually. And with good reason—it is extraordinarily safe to be a passenger on a British train. There were no fatalities in rail accidents for on-board passengers in 2010, or in the preceding three years. Eight passengers were killed in stations: an elderly man falling down an escalator, four people falling off platforms when intoxicated, and so on. This is a rate of 1 death per 170 million passenger journeys, and even then does not reflect the experience of the typical passenger.

Just because there have recently been zero deaths in rail accidents does not mean the future risk is zero, and we can fit a smooth trend to past data, which suggests the risk has been dropping at around 6 percent per year and is now expected to be 1.6 fatalities per year. This corresponds to 33,750 kilometers (20,000 miles) per MicroMort. The UK Rail Safety and Standards Board uses a slightly different method and claims around 12,000 kilometers (7,500 miles) per MM, making trains around 30 times safer than a car per mile. In 2011 in the United States, 6 passengers were killed on all rail for around 38 billion miles traveled, which works out as about 10,600 kilometers (6,600 miles) per MM, similar to the UK rate.[5]

People do fall over on platforms, but the 240 major injuries recorded in the United Kingdom in 2010 work out to less than 1 for each 5 million journeys: the accident rate is higher at off-peak times, when users may be less familiar with the system, or drunk, or both. Maybe that's their fault. Though a guard on a train was jailed for five years in 2012 after signaling for the train to leave while a young girl

leaned drunkenly against it. She fell between the car and the platform and was crushed.[6]

How does the United Kingdom compare with other countries?
From 2004 to 2009 it had the lowest fatality rates in the European
Union except for Sweden and Luxembourg (which has only 170
miles of rail).[7] In contrast, 200 people were killed in just three accidents in 2010 on Indian Railways, although they do carry over 30
million passengers a day. But the United Kingdom has had its own
share of spectacular disasters on the way to safer train travel, such as
a triple collision at Harrow and Wealdstone in 1952 that led to 112
deaths. After World War II the number of passenger deaths regularly
exceeded 50 per year, and around 200 railway workers were also
killed each year: in 2010 the number of workers killed was 1.

But as Huskisson found out, a moving train can be lethal if you
are not in it: 239 people were killed by, rather than on, British trains
in 2010. These were mainly suicides and suspected suicides, but 31
were crossing the line at a legal or illegal place, about half the usual
figure. The number of non-passengers killed has not improved: it
was almost exactly the same in 1952, when 245 died. The number of
suicides has also been amazingly constant, varying between 189 and
233 in the past ten years. Quetelet would have understood. Trains
are even more lethal in the United States—a recent study found that
there were an average of 876 people a year killed by trains between
2000 and 2009: 494 trespassers and bystanders, 281 drivers stuck
on grade crossings, 68 pedestrians at grade crossings, 26 employees,
and just 7 passengers: the proportion of suicides is unknown. Best to
get out of the way when you hear that horn blow.[8]

In Victorian days, a train crash with a few fatalities might only
have made the inside pages of a newspaper. Now the image of a
derailed train would have massive coverage. It is clear that just
counting the bodies fails to measure public interest and concern at
a "disaster." One accident that kills, say, 10 people gets far more attention than 10 accidents that kill 1 each. The 876 people each year
who die on US tracks usually have little publicity, but imagine if
they all occurred at once. . . .

That's a problem if you're deciding whether to spend money on safety, where the government uses the concept of the Value of a Statistical Life (VSL) (or Value for Preventing a Fatality—VPF). As described in Chapter 1, this figure currently stands at around £1.6 million (US$2.7 million) in the United Kingdom, corresponding to £1.60 (US$2.70) to avoid a MicroMort. But the calculation falls apart if lives lost in a group are somehow worth more than lives lost individually. So a multiplier can be used that "up-weights" lives lost in a "disaster" to reflect public concern.

Although railways are generally seen as safe, people still seem to think it a good idea to spend huge amounts of money making them safer. The British Rail Safety and Standards Board employed a consultant philosopher to help them understand why. He concluded that it was a moral issue—people quite reasonably had a sense of shame that train disasters could happen, felt outraged that they were caused by someone's actions, and so didn't tolerate much risk.[9]

But caution can have unintended consequences. After the 9/11 attacks in New York, many people felt more nervous of flying and took to their cars instead. Psychologist Gerd Gigerenzer states that 1,500 more people than usual were killed on US roads over the following year.[10] Similarly, after a train derailment at Hatfield killed 4 people in 2000, speed restrictions were placed on the rail network while the track was checked, and this caused the system to clog up; again people took to their cars, and it has been claimed that an extra 5 or so were killed in the first month (although it is unclear where this figure comes from). Better to stick to the train, as long as you avoid alcohol and escalators—and don't mind pork pies or accents.

ROADS

Roads are unquestionably more dangerous than trains. In 2010 the average annual risk of dying on the roads was about 31 MicroMorts in the United Kingdom, while in the United States the 32,000 highway fatalities from a population of 300 million mean an average of more than 100 MMs per year. It's not that road journeys are more dangerous in the United States, though, it's just that people do a lot more of them, clocking an average of 10,000 miles per year for each of those 300 million people. Since the average US dose of acute fatal risk from external causes is about 450 MMs a year (just over one a day, remember) this means that something like 22 percent of that risk in the United States is on the road. Relative to trains and planes, it is, on average, far more dangerous per mile to drive, ride a motor-cycle, bicycle, or walk wherever there is traffic: you get about 240 miles per MM in a car in the United States, about 90 in a truck, but only about 4 on a motorcycle.[11]

But how many believe they're average? Most think they're bet-ter than that—a self-confidence known, naturally enough, as the "above-average effect" or "illusory superiority."[12] Like everyone else, even Norm once thought he was better than average. This is illusory for obvious reasons. For example, since only half of drivers can be in the top half, if exactly half of all drivers thought they were in the top half, they could in theory *all* be wrong. Even if you're genuinely better, you might not be much better. And if above-average driving ability turns you into an above-average jerk, you might be so cocky that you become a bigger hazard.

Illusory superiority is also linked to an illusion of control: the idea that with the wheel (or handlebars) in our grasp, our fate is in our own capable hands, not at the mercy of the pilot, who loses it and takes the plane down over a poisonous divorce. But although sitting in economy class, trapped, waiting for the wings to fall off, certainly *feels* vulnerable, lack of control evidently does not equal high risk. MB had no control whatsoever during a simple heart op-eration. He didn't say: "Let me do that." Even so, to be busy at the wheel with silken mastery of road and vehicle, fully in control (you think), can give false reassurance and feed self-belief. For risk on the road is clearly determined not only by our own ability but also by that of all the other drunks and lunatics out there, and partly by ac-cidents that have nothing to do with anybody's skill, but are out-of-the-blue hazards, like the stray poodle that runs out.

So again the question: Are we mad? Mad to feel in control of a car when our control is so limited by others; mad to feel more vulnerable in a plane when people far more capable than us are looking after things? One tactic for overcoming fear of flying is to imagine you are at the controls, pulling back the joystick as you take off, like a child pretending. Clearly, we are fooling ourselves. And we know it. We are not that stupid. Perhaps what the pretense does is to remind us that at least somebody is in control, and this is what they will be doing, and we approve.

There is an argument that this double standard—the risk when I'm driving is okay, the risk on a train or plane isn't—is justified even though it seems to reverse the statistical evidence. That argument is backed up by the British Rail Safety and Standards Board's moral philosopher: if others are in control, of course they ought to take more care of me than I might take care of myself. That's their job. If I risk my own neck, it's reckless; if they risk my neck, it's criminal. Hence, runs the argument, there's logic in the "pages 1–12" press coverage of a rail death—someone else is to blame, and the accident is also about public trust, corporate and government responsibility—compared with the "news in brief" treatment for death on the roads, which are more likely to be like fatal DIY, a private misjudgment after which private grief is of little public importance.

You can argue with this, but while we can mock people who prefer the safety of cars to flying, some of this preference is better seen as an expression of trust, or mistrust, than proof of irrationality.

A final complication is that one response to feeling safe at the wheel is to drive more dangerously. This is known as "risk homeostasis," the idea that we have an inbuilt risk thermostat, a level of risk we're prepared to tolerate (or a level of risk that we seek out and try to maintain because that's the level we enjoy). As the risks change—with seat belts, air bags, better brakes—we adapt our behavior to maintain the same level of risk. So, as cars are made safer, many drive faster, and transfer risk to other road users, such as pedestrians, who don't have air bags.

According to John Adams, a specialist in transportation risk, safety would be best achieved with a huge spike on the steering column pointed at the driver's chest and no seat belt. That should recalibrate the risk thermostat of a boy racer or two to about 10 miles per hour. Dudley Moore said the best safety device was a cop in the rear-view mirror. Both in their own way argue that people become more careful the more it hurts not to be. If you want them to drive safely, expose them to danger. The theory of risk homeostasis also works the other way: if we do protect people from the consequences of taking risks, they will take more risks. How this balance of mechanics and psychology pans out is hard to tell in advance. Will the new safety device make more difference to the casualty figures than any change toward riskier behavior that results from people feeling safer?

Adams also argues that one reason why road casualties fall is that some roads get so lethal that no pedestrian will go near them. If this is true, falling mortality on these roads is a measure not of safety but of danger.

In short, the risk of death or injury on the roads is subject to all manner of psychological filters and questions of interpretation. Are the data useless, then? Of course not. They, too, are part of the soup of influences. And in fact, what they seem to say is so stark, the trends so clear, that they can stand a big margin of error or interpretation and still be clear-cut. All data are wrong. The question is whether they're so wrong that you can't draw conclusions from them. Here they are for the United Kingdom.

In 1950, a few years before DS was born, there were 4.4 million vehicles registered in Britain. In 2010 there were eight times as many, which after the growth of the population equals a rise from 1 vehicle for every 11 people to 1 for every 2 people.

A reasonable assumption might be that more vehicles on the roads mean more deaths. Statistics suggest not. There were 5,012 fatal accidents in 1950. By 2010 this number had dropped to 1,850, a fall of 63 percent in absolute terms.

In relative terms the fall is even more extraordinary, given that there was a rapidly rising volume of traffic. For every 100,000 vehicles in 1950 there were 114 deaths. In 2010 there were 5, a 96 percent reduction. When DS remembers how, as a child, he enjoyed riding in the front seat of their old van, which had no regular safety inspection, seat belts, or air bags, and then how people used to drive to the pub, drink all evening, and meander home, he is not surprised. It is a fall from an average in 1950 of 102 MMs per person per year to about 31 MMs per person per year in 2010.

Among car occupants, the number of fatalities is much as it was 60 years ago—the figure was around 20 a week in 1950, rose to 60 a week in the 1960s, and now is back to 20 a week. The main saving of life has been for pedestrians and cyclists: from 60 a week in 1950 (although in 1940 it was even worse, when a staggering 120 a week were killed by unlit vehicles in the black-out against German bombing raids), down by over 82 percent in 2010 to 11 a week. Another way of looking at this figure is that, for every 100,000 vehicles, about 1 pedestrian or cyclist is killed per year. (We will compare this with other countries later.)

These statistics depend on counting bodies, which is grim but easy. Counting accidents and injuries is trickier—What is an injury anyway, and how bad does it have to be to be recorded?—but it's reported that in 1950 there were 167,000 accidents, with 196,000 injuries, and in 2010 almost the same: 154,000 accidents, with 207,000 injuries.

This means that people still crash into each other about 400 times a day, but the proportion of these accidents that are fatal has dropped staggeringly because of speed limits, safety features, improved medical care, and faster access to medical care.

Almost all the richer nations have seen this trend: in the 30 years from 1980 to 2009, road fatalities fell by 55 percent in Australia, 69 percent in France, 63 percent in Britain, 54 percent in Italy, and 58 percent in Spain, in spite of increasing volumes of traffic, although the reduction in the United States was only 34 percent, and deaths actually rose slightly in Greece. For countries that collect the relevant data, we can work out an average number of MicroMorts per 1,000 kilometers traveled: 4 MMs per 1,000 kilometers in the United Kingdom, 7 MMs in the United States, 10 MMs in Belgium, 20 MMs in Korea, 40 MMs in Romania, and 56 MMs in Brazil.[13]

But who bears this risk? In richer countries the majority of road fatalities are the occupants of cars, whereas in poorer countries it's what are known as "vulnerable road users": pedestrians, cyclists, and those whole families squeezed onto a single small motorcycle. In Thailand, 70 percent of all road deaths are riders of two-wheeled vehicles, which will not surprise anyone who has witnessed the traffic in Bangkok.[14]

And it's in low- and middle-income countries that people are exposed to the serious risk. It's estimated that around 1.4 million people are killed on the roads each year, which is around 3,500 a day, and of these, 3,000 are in the developing world—in spite of those countries containing less than half of all cars.[15] The majority of these victims will be vulnerable road users. South Africa sees 15,000 killed each year, a statistic brought into sharp relief when Nelson Mandela's great-granddaughter Zenani was killed in 2010.

The World Health Organization (WHO) predicts that road traffic injuries will go from its current ninth position in the cause-of-death rankings to fifth place by 2030, causing 2.4 million deaths (as well as between 20 million and 50 million injuries), largely of young people, and at enormous cost to the economies.[16] WHO points out that only 47 percent of countries have laws about safety features such as speeding, drinking, seat belts, helmets, and child restraints. And these are often not enforced.

The average risk for an individual from dying on the roads—31 MicroMorts per year in the United Kingdom—is 103 MMs in high-income countries generally, but 205 in low- and middle-income countries. It may seem odd that countries with fewer cars are riskier, but, perhaps surprisingly, deaths per vehicle decline strongly as roads get busier. It is empty roads that are really lethal. This observation even has a name: it is "Smeed's law" of traffic.[17] Smeed's law is reflected in some extraordinary statistics about Ethiopia, which the World Health Organization reports as having only 244,000 registered vehicles in 2007, but nevertheless those vehicles managed to kill 2,517 people. The majority of those who died were pedestrians—that's 1 death per year for every 100 vehicles. The same rate applied to the United Kingdom, with 34 million vehicles, would mean 340,000 people killed each year, instead of under 2,000.

One lesson is that—if we have the money—the bigger a danger becomes through faster and heavier traffic, the more likely we are to do something to control the risks. Risk doesn't just sit there as a settled fact of life, and it can't be separated from how we react to it.

FLYING

"The pilot has advised that there is some turbulence ahead. Please return to your seats and fasten your belts."

As the plane bucks around, the wings flap up and down, babies cry, and the knuckles turn white, is there anyone who remains completely calm, except those who are drunk, drugged, asleep, or all three?

Fear of flying, or aerophobia, is common. Around 3 to 5 percent of the population just won't fly, around 17 percent admit to being "afraid of flying," and 30 to 40 percent have moderate anxiety.[18] We have known risk professionals—archetypes of rationality—who refuse to fly.

It is a treatable condition, and British Airways runs day courses for around £250 that conclude with a 45-minute flight.[19] Sadly they do not publish their success rates, or the counts of those carried gibbering from the plane.

Again, of all the classic fear factors, it is the feeling of complete lack of control over our fate that kicks in hardest when strapped in a plane with the ground a long way off. Perhaps a lack of understanding also contributes—how does it stay up anyway? If we all stopped believing, would it fall from the sky?

And we've seen how the easy availability of negative images influences our perception, and the media certainly like to linger on pictures of wrecked planes. Those of a certain age bring to mind examples from history, from Buddy Holly to the Manchester United soccer team.

But for real aero-disaster porn nothing can beat the unsubtly named Plane Crash Info website, which keeps a running database of all fatal crashes involving commercial airlines, together with lurid photographs and even audio clips from the black box recorders—which are disturbing, to put it mildly. In 2011 it recorded 44 crashes, around 1 a week. Sounds bad, but it's an improvement on 70 in 2001 (including, of course, 4 on 9/11).[20]

Their analysis, vividly illustrated in Figure 19, shows that the cruising part of the flight is by far the safest, taking the most time and resulting in the fewest accidents. Per minute, take-off and landing are around 60 times as dangerous as the middle of the flight. Try muttering this mantra during turbulence.

Plane Crash Info also estimates that human error accounts for over half the accidents—though whether it's pilot error or mechanical failure, there is still nothing you can do about it.

But perhaps you can do something about which airline you fly on: Plane Crash Info estimates that, in the safest 30 airlines, there is a fatal accident, on average, every 11 million flights, and since there

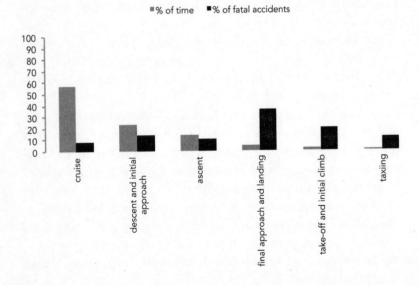

FIGURE 19. What's risky about flying? How the flight-time splits, and when accidents occur, 1959–2008.

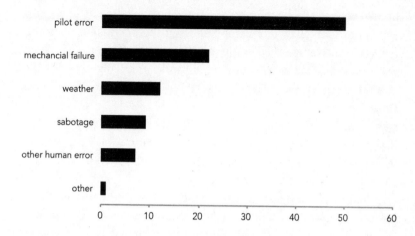

FIGURE 20. Cause of fatal accidents in commercial airlines, 1950–2010.

is some chance of surviving, the odds of being killed are around 1 in 29 million per flight. In contrast, for the bottom 25 airlines, there is around 10 times the rate of fatal accidents, and your chance of being killed on a flight is about 20 times higher.

For calculating MicroMorts, the best data come from the US National Transportation Safety Board statistics on planes carrying at least 10 passengers.[21] Looking at the 10 years from 2002 to 2011, each year US commercial airlines flew an average of 10 million flights. Over that period there were no major disasters, and an average of 16 passengers and crew were killed each year of the 700 million who got on a plane—this works out as 0.02 MMs per flight, or an average of 50 million flights before you are killed. So if you took one flight a day, that's 120,000 years.

Now, you may complain that we have neatly avoided 9/11, in 2001. So if we look at 1992–2011, during which there were a number of major accidents as well as 9/11, it works out at 0.11 MMs per flight, or an average of 9 million flights before you are killed.

But how should we measure these risks so that we can compare them with different ways of traveling? By journey, by mile, or by hour? Let's use as a benchmark the somewhat pessimistic figure of a 1 in 10 million chance of being killed per flight, which is 10 flights per MM. An average US commercial flight lasts 1.8 hours and travels 750 miles. So that works out as 7,500 miles, or 18 hours per MM, around 30 times the distance for driving in the United States, and similar to rail.

As we've seen, risk is not evenly spread through the flight. And if you were really going to choose transportation for a trip on this basis, you would need to factor in the drive to the airport, or the cycle to the railway station, and so on.

But one set of flying statistics stands out from the rest and these are a different matter: small private aircraft, known as "general aviation." There are around 220,000 registered general aviation aircraft in the United States, and in 2012 there were 1,471 accidents, 271 of which were fatal. That's over 5 fatalities a week. On average over the past 10 years, no fewer than 520 people a year were killed in small aircraft and helicopters, 97 percent of all air fatalities.

This works out as 12.4 fatal accidents per 1 million hours flying, around 150 times the hourly rate of commercial airlines. Since generally all the occupants are killed, that's around 1 MM per passenger for every 5 minutes in a small plane: about 12 times the average risk per minute of serving in Afghanistan or Iraq at their worst. At 150 miles per hour, this corresponds to 1 MM every 12 miles, about double the risk, per mile, as walking or cycling.

We see the same pattern in the United Kingdom: between 2000 and 2010 there were 9 fatalities on airlines, 34 in helicopters, and 202 in general aviation.[22] Bigger is definitely safer. So here's a question on the front line of stats versus psychology: Do pilots of small planes feel safer for being at the controls? If you know any, ask.

One last thought. Turbulence can be dangerous even if the wings stay on: 13 people were injured during turbulence in the year up to August 2012 on US commercial airlines. Of these, 12 were cabin crew.[23] So when the seat-belt light goes on, buckle yourself in and just be pleased you are a passenger and not pushing the drinks trolley.

16

EXTREME SPORTS

A slow drive up eleven hairpin bends on the road known as the Troll's Ladder, then a slower climb on foot following the cairns, avoiding loose rock, led to the giant, conical mountaintop called Bispen, "the Bishop," standing like an upturned ice cream cone.

Here the air was still, clear, cold. Far below, far away, the lake was a hard spangle of light between hills and granite. High, thin clouds watched the valley. The world of people disappeared. Here was nothing but view and drop, landscape, distant shapes and textures, shades of gray and green.

On the edge stood Kelvin, alone and as scared as anyone else on a mountain edge.

He jumped.

From above, he was a wrinkle plunging down. From below, he was a stick man etched against blue. In his own space, Kelvin had stretched out the webbed arms and legs of his flying suit against a riot of noise—and was free.

"You just kind of go wherever you look," they said about base-jumping from mountains in a wingsuit, "just kind of . . . stick your arms out."

At near-terminal velocity he planed out, swooped into a glide. Last time, he "cleared out the space," as they put it, got the hell out of there and flew away from the mountainside. But that got boring. He tilted his shoulders toward the cliff, measuring his line past the vicious hard road of the rock face over flashing shrubs and bare pronged trees, past jots of reflected light and sudden shadows as the land rose and fell in spikes and bumps and hollows and crevices, and beyond them all in roaring buffeting air he curved and scored the long, smooth line down.*

Buzzing the walls, they called it, hands outstretched, a fingernail from pieces. "Mess up and you will 100 percent die," they said. "Click. Like that." Kelvin knew it: knew it, that is, as well as any of us knows that our own death is real.

"Sick, man, real sick," they'd be saying in a, like, totally envy-racked and mind-blown way as Kelvin bombed past the watching crowd on a Trollsteigen hairpin at more than 100 miles an hour, here and gone, a whistling projectile so vicious that he could smash a hole in your house but was, by the way, a human being. Cool.

For another minute he roared above forest, arms embracing vast air. Not really flying, "falling with style," they said. "The most extreme badass activity on the planet."

And as the lake drew closer, Kelvin pulled in his arms, brought his heels together, and stood up, still blazing over the water, but slowing now he pulled his cord and released the chute. And then, violently, there was peace. Only the gentle sound of Kelvin's laughter and bottomless exhilaration, the topless high as he steered toward the shore.

And as he drifted down, he thought of those who said this was impossible. Guess what? He just did it. He thought of Graham Greene playing Russian roulette with his father's revolver, and some memory he scarcely recognized as his own about Dostoevsky and a piano.

"Sick! Kelvin man!" They said, as he walked back with his chute. "Dude!"

"Sick!"

---

* For the most jaw-dropping, eye-popping, vicarious risk-taking, watching wingsuit flying is hard to beat. Try this video: "The Best Wing Suit / Skydive from YouTube," Part 1, 2008, www.youtube.com/watch?v=5N9t5qOSzCU&feature=youtube_gdata _player.

"Yo!"

"Okay!"

No doubt about it: near death was real living.

. . . .

THERE IS A BIANNUAL competition for jumping off cliffs in a wing-suit. It is, incredibly, a race. The aim is to be fastest down. It is called the Base Race, organized by the perfectly named (for a risk-taker) Paul Fortun from Germany. You'd think, unlike Kelvin, he'd be fearless. How else could he step off a mountain? But in 2012 he said: "I'm terrified every time I jump, everyone is. I think if you're not scared of something like this, then why bother?"[1]

Which is a neat inversion of the precautionary principle—do it *because* it might go wrong. "It's all about the taste of fear and lack of control," he said. "I love it."

We've discussed mortal danger as if it's a bad thing. But danger can be a scream. Oddly, those who love it and those who hate the idea needn't necessarily disagree about the risks. Paul Fortun and Kelvin do it not because their perception of the probability of death is different from other people's, but because it's much the same. That's why they like it.*

Even when life was brutish and short, with daily risk from disease, hunger, and conflict, and you might guess that not many would choose to stand, or jump, sportingly in harm's way—there being enough harm around already—people boxed bare-knuckle from the early sixteenth century, played, or more accurately, fought, the kind of mass soccer slugfest without referees or rules that occasionally left people dead and maimed, or, if rich, jousted (in medieval times) and (later) galloped full-tilt over hedges.

As life became cozier and more predictable for richer citizens of nineteenth-century Europe, there was the Alpine Club—membership

---

* Although this is unusual. On the whole, if people like the benefits of risk-taking, they tend to think the risk is objectively lower, even if still high enough to be thrilling. See index for "affect heuristic."

criterion: "a reasonable number of respectable peaks"—founded in London in 1857, when the British dominated mountaineering and bagged the classics in desperately unsuitable clothing. These gentlemen—and one or two women—had many motivations. For some it was science, for others a spiritual union with nature, for a few "the cool keen finger of danger."[2]

Climbing mountains was risky then and can be risky now. Apart from the chance of falls, there is low oxygen, freezing temperatures, exposure to wind and sun, and exertion. But it's difficult to put numbers on these risks. You can count dead bodies. But as a proportion of all who go climbing? Or of only those who attempt high peaks per day of activity? Another problem is that, whatever the measure, we need to know how many people get up to whatever it is that turns them on, and these data can be hard to find. So the following numbers are bound to be rough and ready.

By the end of 2011, 219 people were known to have died climbing Mount Everest, 1 for every 25 who reached the summit, since records started being kept in the 1920s. Of the 20,000 mountaineers thought to be climbing above 8,000 meters (26,000 feet) in the Himalaya between 1990 and 2006, it was estimated that 238 died, a rate of around 12,000 MicroMorts per climb. In another study of 533 mountaineers on British expeditions above 7,000 meters (23,000 feet) between 1968 and 1987, there were 23 deaths (1 in 23), equal to about 43,000 MMs per climb. Mountaineering at these levels is riskier than a bombing mission in World War II, or about equal to 117 years of average acute risk.[3]

Next, falling with style. For while some are terrified at the thought of stepping into an airplane, others enjoy stepping out midflight, or otherwise hurling themselves into the air. The dangers have been apparent since Icarus first strapped on fake wings and discovered that the birds had made it look too easy.

Parachutes were developed in the late 1700s, hang gliders 100 years later. This was a time when inventors were expected to demonstrate their bolder ideas personally. One such was Franz Reichelt, in 1912. He was a tailor who invented a wearable parachute, like a cross between a voluminous raincoat and an inflatable dinghy. It worked with dummies.

He assured the authorities that it was a dummy he wanted to test at the Eiffel Tower. At the last moment he clambered on top of the railings wearing the parachute and could not be dissuaded. Early film shows the heart-stopping moments as he rocks back and forth, finding the courage to jump, then flutters forlornly down like a shot bird. The crowd, silent in the film but doubtless screaming, rush forward to see his body on the frozen gravel. The police are then seen measuring the depth of the dents.[4]

Parachuting competitions began in the 1930s, and the US Parachute Association estimated that an average of 2.6 million jumps were made each year between 2000 and 2010.[5] But it's still not entirely without risk: in that period 279 people died, around 25 a year, for a rate of about 10 MicroMorts per jump. Examination of the details of the accidents reveals these were mainly experienced enthusiasts with thousands of jumps behind them, pushing things a bit far. Novices were safer.

Reichelt could be considered one of the first base-jumpers, a sport in which skydivers leap from fixed objects rather than planes. This, it hardly needs saying, is dangerous, although, perhaps surprisingly, it can be less dangerous to jump from certain mountains than to climb an especially tall one. The Kjerag Massif in Norway is one of the safer launch points, with a sheer 1,000-meter drop. In theory, this allows ample time to do something to soften that inevitable reunion with the ground. Nevertheless, over 11 years and 20,850 jumps, there were 9 deaths and 82 nonfatal accidents: that's 1 death in every 2,300 jumps, or an average of around 430 MMs per jump.[6] And this is one of the safer spots: base-jumping is not a mass-participation sport, for obvious reasons, but 180 deaths have been recorded, increasingly featuring wingsuits that open up like a flying squirrel, much as poor Reichelt originally intended.

If diving through the air doesn't appeal, there's always diving underwater. Jacques Cousteau's development of the "Aqua-Lung" in 1943 turned scuba into a leisure activity, and the British Sub-Aqua

Club (BSAC) now has more than 35,000 members. The BSAC keeps a careful tally of diving fatalities and recorded 197 deaths in the 12 years from 1998 to 2009, an average of around 16 a year.[7] It estimated about 30 million dives over this period, so the average lethal risk was around 8 MMs per dive. But this is an average—for BSAC members it was 5 per dive, for nonmembers 10.

Diving, like modern mountaineering and parachuting, relies on technical developments to reduce the risk of an inherently dangerous pastime. By comparison, running seems natural and moderate, one for the more cautious. But long-distance running, as Pheidippides found 2,500 years ago, after collapsing and dying from his run of 26 miles to announce victory at the Battle of Marathon, can end in more than sore feet. In 3.3 million marathon attempts in the United States between 1975 and 2004, there were 26 sudden deaths. That's around 7 MMs a marathon on average, roughly like a scuba dive or a skydive. Recent attention has focused on the risks of drinking too much water, from which one participant in the London Marathon died in 2007.[8]

Most weekend sports aren't likely to kill you, but injuries are common. In the United States, the National Electronic Injury Surveillance System (NEISS) produces a vast catalog of just about every sporting misadventure you could imagine, and a lot of other kinds, too. Based on a sample of hospitals, all life passes through these data. Search for "football" and you discover a "national injury estimate" (we stress the "estimate") of 466,492, which you can dip into, to see, for example, a record for January 2, 2012, of a 17-year-old male with a sprain of the toe caused by stepping in a hole on a football field. Or 66,543 estimated cases (from a sample of 1,652) of horse-riding accidents, including the 72-year-old who fell off and fractured her pelvis; 460,000 accidents with exercise equipment; and 114,120 skateboard accidents (an estimated 16 of them happening to over-65s).[9]

In the United Kingdom, too, data used to be collected on a sample of people treated in hospitals for accidents from leisure activities, and the resulting Leisure Accident Surveillance System (LASS) statistics recorded, for example, that 620 people were injured in riding schools in 2002.[10]

But this was based on only around 17 of the hospital accident departments in the United Kingdom. Scaled up to the whole country, the estimate is of about 12,700 accidents at riding schools. Similarly, LASS estimates 6,500 accidents at golf clubs, and nearly 700,000 injuries from all sports, of which the majority—450,000—were from ball games, enough to make you want to stay home and watch soccer on television. LASS even identifies the specific object associated with each injury: 200 from javelins (best not to ask), 1,600 from skipping ropes, 17,000 from cricket balls, 260,000 from soccer balls, 34,000 from skateboards and rollerblades, 3,200 from fishhooks, and 5,800 from bouncy castles.

MB asked a budding *traceur*—a practitioner of parkour, or freerunning—about the injury risks of rolling, jumping, and vaulting up, down, and over urban obstacles at speed, and she rebuked him for not knowing a technique for avoiding risk by moving swiftly but safely when he saw one. To which you can't help replying, if safety is the point, why not take the stairs normally? (Though stairs have hazards, too: see Chapter 18, on health and safety.)

Clearly there's more to it than being careful. Equally, most of the recreational risk-takers don't want to die. Taking skydiving, scuba diving, marathon running, and other moderately extreme sports, there seems to be some natural level of risk—up to around 10 MicroMorts—that most participants are prepared to accept while remaining reasonably sensible. This does not include base-jumping or climbing to high altitudes.

So it is not illogical for people to talk of safety even as they flirt with danger. Dangerous sports do not equal recklessness, for the risk seems to be carefully, if not always consciously, calibrated, and strangely consistent across a wide range of activities. The acute risks of these activities are certainly higher than those of an ordinary day at the desk, but for most participants of most extreme-ish sports, the risk of ten times the average daily dose seems to be about where their risk thermostat is set on the weekend.

One study by Stephen Lyng of what he calls "edgework," a phrase coined by Hunter S. Thompson in *Fear and Loathing in Las Vegas* (see also Chapter 9, on illegal drug-taking), describes edgework, in the bracing language of social theory, as "acquiring and using finely honed skills and experiencing intense sensations of self-determination and control, thus providing an escape from the structural conditions supporting alienation and oversocialization."[11]

Hate the job, feel like a cog in the machine, tired of the daily MicroMort? Want to feel spontaneous and forget your everyday self? Then squeeze about ten days normal risk into one—and jump.

As with drugs, risk perception is capable of turning every threat into an attraction. But it usually stops short of making death desirable. Some speak of a lack of control, others of the fine judgment that enables them, just, to keep control. Either way, they think they will survive to tell the tale. And since they always have the option of leaving the danger behind to do something else, the big element of control that they all retain is the freedom to choose whether or not to do it. To take that point to absurdity, Kelvin would find no thrill in extreme proximity to death in old age. Hardened arteries somehow don't count as edgework. So choice is a big factor in attitudes to danger. The fright factor of involuntary risk, or risk that people can't escape—exposure to pollution is another good example of a risk we can't avoid—often feels far worse than risks we choose, such as adventurous sports.[12] Surfers Against Sewage is a real organization, with a fair point about which risks are tolerable. They don't mind the chance of being wiped out, drowned, or knocked on the head by a surfboard so much as they mind swimming in effluent. And who can blame them? This isn't just about avoiding danger. Base-Jumpers Against Climate Change—if it existed, and depending on whether you believe in it—wouldn't be altogether daft either.

If parachuting were as safe as a day at the office, would it have the same thrill? For many, yes. The allure is not the risk as it appears in the statistics, but the risk as it feels in the moment. Instinctive fear of being high up is probably an evolutionary plus. But evolution hasn't caught up with parachutes yet. In other words, danger isn't measured only in the head, it's measured in the skin. As with flying or roller coasters, we can feel afraid, even enjoy the fear, with only a little more than usual to worry about.

Objective measures of harm are not irrelevant to this subjective thrill. If you knew the fairground ride was a death trap, you wouldn't get on. So we say to ourselves, "I know I'll be all right, but I don't feel it," and with that careful balance enjoy the ride. For many others an old saying has to be taken seriously: risk is its own reward. Just don't be surprised when people choose different levels of risk for different aspects of their lives. It's not illogical. It's not necessarily inconsistent. Attitudes to danger vary even within individuals, for any of whom risk is strangely but carefully compartmentalized.

LIFESTYLE

Norm lay in bed, eyes shut, listening to his body. He was middle-aged and felt it. His body told him that it hurt.

He had recently taken up exercise after studying the data on chronic illness over lunch with Kelvin: a burger for Kelv, a salad for Norm. Running was his attempt to slow the raindrop as it trickled down the window. Although there were other consolations, too. For example, beer tasted better afterward, bitter with the sweet taste of righteousness.

So for twenty minutes each day on 200 milligrams of flecainide acetate to control his cardiac arrhythmia, paunch creeping over the band of his shorts, he forced his feet down lanes lined with brambles or through damp leaves by the canal, blowing hard at the lid of his closing coffin, then sucked the air back like a glutton. He heaved his carcass all the way up to jog speed, nearly, until the aches were washed out by adrenaline. This was what it was all about. This was striving.

"Commit, Norm, commit!"

He pushed again until his body screamed that it was alive and feeling too much, all to counter the long hours when it seemed to feel nothing but fed up. He drove his thin legs and squeezed wheezing lungs to prove his frame could raise one more militant shout against age and wouldn't be dried and silenced. When he was running, he was living; it was transcendent and forever, for 22 minutes.

And Norm knew precisely how much extra life 22 minutes of exercise was worth, on average.

He finished. He slumped, while every muscle said it was spent, and he wondered if he could have sustained his Usain Bolt finishing burst from the gate into the allotment instead of waiting for the lane. And he looked at his watch and he saw—and sighed—that he was, after all, at 22 minutes and 18 seconds, slower. Now he could go to the pub.

. . . .

LIFESTYLE IS A new kind of danger to Norm. The hazards he's lived with so far have been instant, like violence or accidents, the kind that hit us over the head with a swift goodnight. But Norm is now running from a more sinister threat, another type of mortal hazard with slower effects that go stealthily into the blood one cancerous bacon sandwich or poisonous drink at a time, potential killers by degrees that might catch up with us later in life, as something surely will.

The first mortal hazard—the quick one—is known as acute risk; the second is chronic. Murder with a chainsaw is an acute risk, obesity a chronic one that takes time to do its worst. Of course, the same hazard might be both: too much booze can do you in quickly when you fall under a bus, or slowly stew your liver. But in general it helps to separate them.

So far, we've used the MicroMort to describe acute risks. For chronic risks, such as obesity or Norm's current long-term lifestyle anxieties, we introduce a little device we have called the MicroLife (ML). Here's how it works.

Imagine the duration of your adult life divided into 1 million equal parts. A MicroLife is one of these parts and lasts 30 minutes. It is based on the idea that as young adults we typically have about 1 million half-hours left to live, on average.*

---

* Life-expectancy for a man aged 22 in the United Kingdom is currently about 79, or another 57 years, which is 3,418,560 minutes, or 20,800 days, or 500,000

Sounds unimpressive. But we like the MicroLife. It is a revelatory little thing. Like the MicroMort, it brings life down to a micro-level that's easy to think about and compare: half-hour chunks, of which we have 48 a day. Think of it as your stock of life to use up any way you choose, 1 million micro-bits of a whole adult life, each worth half an hour, yours to spend. Watching the World Series? Bang, 6 MicroLives gone, just like that, never to have again.

So the simple passing of time uses up MicroLives. Every day we get up, move around, stuff tasty things into our bodies, discharge smelly things out of our bodies, and go to bed—perhaps with the thought, if we're gloomy, that there go another 48 MicroLives from our allotted span.

But extra MicroLives can also be used up by taking chronic risks. So although time passes to its own beat, our bodies can age faster or slower according to how we treat them. If we jump around more, and stuff less or better, how much can we slow the steady tick-tock toward disease, decrepitude, and death? And if we indulge and allow ourselves to be couch potatoes, how fast might our own clocks run?

In other words, MicroLives can measure how fast you are using up your stock of life, faster or slower depending on the chronic risks to which you're exposed. If your lifestyle is chronically unhealthy, you'll probably burn up your allotted MicroLives that much quicker, and die sooner, on average.

For example, lung cancer or heart disease often follows a lifetime of smoking, and subsequently reduces life-expectancy—again not for everyone, but overall. Some people seem indestructible, smoke like a chimney, and drink like a fish, and never look the worse for it. But, on average, even if chronic risks don't kill you straight away,

---

hours, or 1 million half-hours. For a woman, life-expectancy is about 83, so her million half-hours start at 26. You can take 2 years off for the United States, so that 20-year-old men and 24-year-old women expect to have 1 million half-hours ahead of them, ignoring the probable improvements that will happen in their lifetimes. This will not be true for everyone, but short of clairvoyance it'll do, roughly, more or less, overall.

they tend to kill you sooner than if you had avoided them. Again, if we count the bodies, we can estimate how many years are lost overall, whether to obesity, smoking, or sausages, and we can convert this loss of life into the number of MicroLives burned up by unhealthy living. Thus, exposure to a chronic risk equal to 1 ML shortens life, on average, by just one of the million half-hours that people have left as they enter adulthood.

It turns out that one cigarette reduces life-expectancy by around 15 minutes, on average, and so two cigarettes cost half an hour, or 1 ML. Four cigarettes are equal to 2 MLs.

The first two pints of strongish beer also equal about 1 ML. Each extra inch on your waistline costs you around 1 ML every day, 7 a week, about 30 a month and so on. According to recent research, so does watching two hours of TV. An extra burger a day is also about 1 ML. We'll reveal the calculations behind these risks in a moment.

We could simply add up all these MicroLives, half-hour by half-hour, to see roughly how much time, on average, you lose in total from whatever your life span might have been. But the end of life is often far away, like the end of the story, and a lost half-hour deferred until you are in your dotage hardly seems to count. As a doctor said in a British newspaper: "I would rather have the occasional bacon sarnie than be 110 and dribbling into my All-Bran." But by thinking of exposure to chronic risks like an acceleration of the speed at which you use up your daily allotment of MicroLives, we can do something more vivid and immediate. We can show how much your body ages each day according to the chronic or lifestyle risks that you take.

Ordinarily, remember, we use up 48 MLs a day. But remember, too, that smoking 4 cigarettes burns an extra 2 MLs. So if you smoke 4 cigarettes in a day, you've used not 48, but 50 MLs that day. In other words, after a 24-hour, 4-fag (to use the British slang) day, we could say that you are 25 hours older.

And that's not a bad representation of what can happen biologically. Bodies do often age faster when we do bad stuff to them. Twenty cigarettes daily means you burn an extra 10 MLs a day, on average, or become 29 hours older with every 24 that pass, or move toward death 5 hours faster, every day.

Suddenly, chronic risk feels a lot more here and now than the faraway payback that we typically put off thinking about until we're spent anyway. It counters what is sometimes known as "temporal discounting." By bringing chronic risk down to the small scale of what happens today, rather than thinking of it only as a life-size problem deferred, the MicroLife makes chronic risk a good deal more real and immediate.

But should it? You might argue with this. You might prefer to put off facing up to your lifestyle risks. You might argue that you shouldn't be confronted with the payback until it actually occurs, late in life. On the other hand, it could be argued that the damage is done now, at the point of consumption, so we should measure it now.

MicroLives allow us to make simple comparisons of chronic risk, just as we did for acute risk with MicroMorts. Now we can compare sausages with drinking or smoking, X-rays with mobile phones. We can compare a CT scan with watching the Hiroshima atomic bomb from the Hiroshima suburbs, getting fat with getting fit, unprotected sex with unprotected sun—and we will. In Chapter 27, Figure 38 shows a selection of MicroLife hazards.

If calculations terrify you, move swiftly on to the next chapter. Because next, we'll find out exactly how these calculations are done and what the evidence is for the MicroLife costs and benefits we've identified. Treat it as a statistical detective story.

We begin with the 1 ML cost of an extra burger every day, mentioned earlier. This was reported in the British newspaper the *Daily Express* in a story about the dangers of red meat, based on a study from Harvard University.[1] The *Express* said: "If people cut down the amount of red meat they ate—say from steaks and beefburgers—to less than half a serving a day, 10 percent of all deaths could be avoided."

Oh to be one of the 10 percent for whom death could be avoided! But this is not what the study said. Its main conclusion was that an extra portion of red meat a day—this being a lump of meat around the size of a pack of cards or slightly smaller than a standard quarter-pound burger*—is associated with a "hazard ratio" of 1.13: that is, a 13 percent increased risk of death. Put aside any doubts about the validity of this number for a moment and take it at face value. What does it mean? When our risk of death is already 100 percent, surely a risk of 113 percent is an exaggeration?

Let's consider two friends—whom we'll call Kelvin and Norm— who are both aged 40, and just for the moment let's make the unrealistic assumption that they are pretty much alike in most lifestyle respects that matter, apart from the amount of meat they eat.†

Carnivorous Kelv eats a quarter-pound burger for lunch from Monday to Friday, while Normal Norm does not eat meat for weekday lunches, but otherwise has a similar diet to Kelvin's. We are not concerned here with their friend Particular Pru, who has given up eating meat and turned veggie after reading the Daily Express, but who might succumb to a contaminated sprouted fenugreek seed.

Each person faces an annual risk of death, the technical name for which is their "hazard," or, somewhat archaically but poetically, their "force of mortality" (for a fuller discussion of the force of mortality, see Chapter 26, "The End"). A "hazard ratio" of 1.13 means that, for two people like Kelvin and Norm, similar apart from the extra meat, the one with the risk factor—Kelvin—has a 13 percent increased annual risk of death—not an overall risk, obviously— during a follow-up period of around 20 years.

This does not imply that his life will be 13 percent shorter. To work out what it really means we have to go to the life tables provided by the US Centers for Disease Control and Prevention. These tell us the risk that an average man—Norm, say—will die at each year of age. In 2008 this risk, or hazard, was at its lowest for those aged 10 (see Chapter 2, on infancy), at less than 1 in 10,000: it then

---

* Actually 85 grams, or about 3 ounces.

† The assumption is—in line with the Harvard study—that they have the same average weight, alcohol intake, exercise regime, and family history of disease, but not necessarily quite the same income, education, and standard of living. This is how the Harvard team analyzed the risks, by trying to focus on the effect of the meat we eat without too many other factors in the way.

rises to 1 in 1,000 at age 19, then to 1 in 100 at age 59, until, at age 85, 1 in 10 will die before their 86th birthday. Very roughly, the annual chance of death increases tenfold around every 27 years, which works out at doubling every 9 years, or about a 9 percent extra risk of dying before the next 12 months are out, for every year that we are older. The tables also tell us life-expectancy at any given age, assuming the current hazards, and, having survived to age 40, Norm is expected to live another 38 years, until he is 78.[2]

From this we can work out Carnivorous Kelv's prospects by multiplying all Norm's hazards by 1.13. After a little work in a spreadsheet, we find that Kelvin can expect to live 37 more years, on average, a year less than Norm. So Kelvin's lunch—if he eats the same lunch all his life and if we believe this hazard ratio—is associated with the loss of one year in expected age at death, from 78 to 77.

Is that a lot? Kingsley Amis, the English novelist and poet, said, "No pleasure is worth giving up for the sake of two more years in a geriatric home at Weston-super-Mare" (you do not need to have an intimate knowledge of Weston-super-Mare to get the flavor of this comment).[3] It is for readers to decide. But we cannot say that precisely this amount of time will be lost. We cannot even be very confident that Kelvin will die first. In fact, there is only a 53 percent chance* that Kelvin will die before Norm, rather than 50:50 if they eat the same lunch. Not a big effect.

But it sounds rather more important if we say that this lost year (around 1/40th of the remaining life) translates very roughly to one week a year, or roughly half an hour a day: that's one extra MicroLife burned up for each daily burger. So, unless you're a very slow eater, you expect to lose more life than the time it takes to eat your burger.

But we can't even say the meat is directly causing the loss in life-expectancy, in the sense that, if Kelvin changed his lunch habits

---

* If we assume a hazard ratio h is kept up throughout their lives, then some rather elegant math tells us that the probability that Kelvin dies before Norm is precisely $h/(1+h)$, which, when $h = 1.13$, means a 53 percent chance Kelvin dies first, rather than 50:50 if they eat the same amount of red meat.

and stopped shoveling down the burgers, his life-expectancy would definitely increase. Maybe there's some other factor that both encourages Kelvin to eat more meat and leads to a shorter life.

Income could be such a factor—poorer people in the United States tend to eat more burgers and also live shorter lives, even allowing for measurable risk factors. But the Harvard study does not adjust for income, arguing that the people in the study—health professionals and nurses—are broadly doing the same job. We think that many of these studies about diet should be taken with a pinch of salt (although perhaps not too much, since it may increase your risk of heart disease).

We can also look deeper into the calculation for other bad behaviors that receive a finger-wagging. So, next, smoking. The evidence against smoking is much better than that against red meat. In terms of shortening your life, a very basic analysis estimated a 6.5-year difference in life-expectancy between smokers and nonsmokers, which is 3,418,560 minutes. They considered median consumption of 16 cigarettes a day from ages 17 to 71, which comes to 311,688

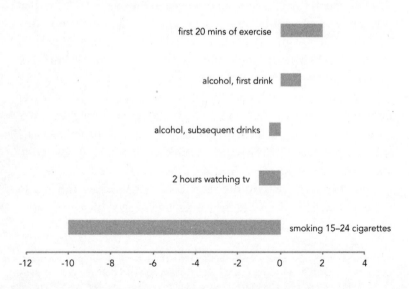

FIGURE 21. Some MicroLives. How many gained or lost for various activities.

cigarettes. Making the simplifying assumption that each cigarette contributes equally to the risk, this comes to an 11-minute loss in life-expectancy per cigarette, or around 3 cigarettes for a MicroLife.[4]

This simple analysis only compares smokers with nonsmokers, who may be different in all sorts of other ways that could also influence their health. A more sophisticated analysis would consider the effect of stopping smoking, and fortunately that's been done in a classic study on 40,000 UK doctors, many of whom gave up smoking during the period of the study from 1951 to 2001.[5] They estimated that a 40-year-old man who stopped smoking gained 9 years in life-expectancy, or 78,000 hours in total, and, if we assume the effects are roughly proportional, from this we can estimate around 2 cigarettes for a MicroLife. So 2 ciggies is roughly equivalent to 1 burger, when taken daily.*

So what about the booze? The precise effect of alcohol on all-cause mortality is controversial, since, although it can cause accidents (particularly for drivers and young binge-drinkers), give you liver disease, and increase the risk of some cancers, it can also protect your heart. So the "dose-response" curve for annual risk is J-shaped in middle age, meaning that the risk falls slightly with the curve of the "J" as you consume a small amount of alcohol, then rises as you consume more. Very roughly, it looks like the first drink each day adds a MicroLife, but extra drinks take it away again, and more. So the first is medicine, the second poison, the third poison, and so on. It does not go medicine/poison/medicine/poison . . . —that would be ridiculous, and anyway would mean that one would need to have an odd number of drinks.[6]

All this is depressing, but what about the benefits of a good diet and hearty exercise—eating muesli and running (although not at

---

* If he carries on smoking, he is only expected to live another 30 years, or 11,000 days, so he is, on average, losing 7.2 hours per day (14 MicroLives): we can imagine him speeding recklessly toward his death at 31 hours a day. During these 30 years he might smoke 325,000 cigarettes (assuming the higher consumption of 30 a day in the 1950s and 1960s). This works out to 15 minutes lost per cigarette smoked.

the same time)? As A. A. Milne, author of the Winnie-the-Pooh stories, wrote, "A bear, however hard he tries, grows tubby without exercise," and we can try and work out what that tubbiness might do to your life. A recent review estimated a risk that translates to around 1 ML lost for every day that you are 5 kilograms (about 11 pounds) overweight.[7] Seriously obese people can lose 10 years off their life-expectancy, similar to smokers.*

The European Prospective Investigation into Cancer and Nutrition (EPIC) study in Norfolk in the United Kingdom compared people who ate their five portions of fruit and veggies a day with those who didn't (and the researchers checked their honesty about their diet by measuring vitamin C levels in their blood).[8] The hazard ratio was 0.69, showing around 30 percent less annual risk associated with a better diet, or around 3 ML a day saved, as if they were only aging 22.5 hours instead of 24.

And all that running around? Both the US and the UK guidelines recommend we all get 30 minutes of moderate or vigorous activity five days each week—a total of 2.5 hours per week, or 22 minutes a day. When asked in 2008 about their physical activity, 39 percent of UK men and 29 percent of women claimed they did this much exercise each week, while in a phone survey, an impressive, and possibly implausible, 52 percent of US adults fulfilled the recommendations.[9] However, exercise, like sex, tends to be overreported, just as alcohol intake is underreported—if we believed what people say they drank, half the alcohol bought in the United Kingdom must be left in the bottle or chucked down the sink. Which we doubt. When UK adults carry accelerometers that measure their true exercise, only 6 percent of men and 4 percent of women meet the government-recommended levels of activity. We are part of a population that is both tubby *and* deluded.

---

* The study gives an estimated hazard ratio of 1.29 for all-cause mortality per 5 kg/m^2 increase in body mass index (BMI) over the optimum of 22.5 to 25 kg/m^2. For a man or woman of average height (1.75 meters and 1.62 meters, respectively, or about 5 feet 9 inches for men and 5 feet 4 inches for women), this corresponds to a hazard ratio of around 1.09 per 5 kg (again, about 11 pounds) overweight, translating to one MicroLife per day.

A huge review of 22 studies involving nearly 1 million people concluded that 2.5 hours a week of "non-vigorous" activity was associated with a hazard ratio of 0.81 compared with being a complete couch potato—producing a 19 percent reduction in annual risk of death.[10] This works out as about an hour per day, or 2 MLs, added to life-expectancy for an average of 22-minutes-a-day activity, which must be why that's the length of Norm's run—quite a good return for the investment of getting off the sofa.

So it's a good idea not to be a couch potato, and a Swedish study even showed that it's never too late to start, with increasing physical activity in middle age eventually bringing your risk down to the level of people who have always been active—similar to the benefits of giving up smoking.[11]

Some naïve extrapolation would suggest that if we do lots of exercise we could live forever, but sadly, there is a strong law of diminishing returns. People who do 7 hours of moderate exercise a week, a full hour per day, only reduced their risk by 24 percent, equivalent to about 1.5 hours added to their life for each day of exercise.

So, very roughly, compared with doing no physical activity, the first 20 minutes or so per day pays rich dividends, but any more than that is approximately pro rata: 20 minutes of exercise earns you 20 minutes of added life-expectancy. It's as if time stops for you when you're exercising. And on a treadmill it can certainly feel that way.

Our circumstances when we're born—which we can't do much about—can also be expressed in MicroLives. For example, being female rather than male (worth an extra 4 MLs, or 2 hours per day), being Swedish rather than Russian (21 MLs a day for males, or more than 10 extra hours), and living in 2010 rather than 1910 (15 a day, or 7.5 extra hours).

Of course, working out the health effects of lifestyle is not a precise art. It is impossible to put exact numbers on the harm we do to our bodies with an extra cigarette, sausage, pint of beer, or glass of wine, or the lack of five-a-day fruit and vegetables, or—on the virtuous side—the benefits we might reap on the exercise bike or by being otherwise wholesome and pure.

But we can calculate the effects approximately, by averaging over many lifetimes, and it is worth trying, especially in a world of exhortations to self-improvement or stories of how you, too, can stay forever beautiful and young—or not.

There is one big difference between MicroMorts and MicroLives. If you survive your motorcycle ride, your MicroMort slate is wiped clean and you start the next day with an empty account. But if you smoke all day and live on pork pies, then your used-up MicroLives accumulate. It's like a lottery where the tickets you buy each day remain valid forever—and so your chances of winning increase. Except that in this case, you really don't want to.

People spend a fortune on cigarettes, but what would you sell half an hour of your life-expectancy for? Governments put a value on MicroLives, as they do with MicroMorts. The UK National Institute for Health and Clinical Excellence (NICE) has guidelines that suggest the National Health Service will pay up to £30,000 for a treatment that is expected to prolong life by one healthy year. That's around 17,500 MLs. This means that NICE prices a MicroLife, or half an hour of your remaining life-expectancy, at around £1.70 (US$2.80), almost exactly what the UK Department of Transport says it would pay to avoid a MicroMort.

Does this mean the government ought to pay you a few dollars for every two cigarettes you resist, or for every day that you keep that extra inch off your waistline? Possibly, except that it wouldn't work like that: we could all claim to resist hundreds of cigarettes a day, nonsmokers especially.

Now that Norm is into middle age and full of regret for the burgers that slipped through in earlier life,* and he is puffing hard to put time back in the bank, it's interesting to note that he hasn't become pure, merely smug. A 22-minute run might buy him an extra hour of life, but one effect is that he feels happier about spending it in the pub.

---

* Kelvin? *Il ne regrette rien.*

FIGURE 22. Scandinavian orange juice. But who would you talk to?

His behavior is known as risk compensation, related to the idea in Chapter 15, on transportation, that we all have a built-in risk thermostat. Taken your vitamins? Great. Eat extra fries!

There's experimental evidence to support this. In one trial, in a culture of heavy smokers, some participants were given pills and told they could have a cigarette break. All the pills were placebos. But those who thought they'd taken a vitamin pill were far more likely to go for a smoke (89 percent, compared to 62 percent). In another study surveying people's sense of vulnerability, those who took vitamin pills somehow thought they were less likely to be hurt in an accident.[12]

So if you've banked some health, you might feel free to spend it. How these attitudes and behaviors shake out—some beneficial, some compensatory and harmful—takes some calculating, except for vitamin pills, which don't do most people much good anyway. But it does suggest that the risk calculation for healthy behavior also needs to take account of the cream cake afterward.

A summary of the research on exercise suggests that it still brings substantial benefits, although less weight loss than you might hope, perhaps because people eat more.[13] The research does not suggest that exercise leads people to eat so much more that they become fatter on average, contrary to much online gossip. But do mind that you don't seek too much cake compensation for all your hard work.

18

HEALTH AND SAFETY

Nearly home, Norm saw a line of plastic bollards outside his house where the road swerved up and over a small hill. At either end was a traffic light, in the middle a van and generator. On the back of the van was written "Urgent Response Vehicle." Beneath this: "Limited to 56 m.p.h." Next to the van stood a man studying a plan.

"What's up?" said Norm.

"Two-way traffic control, sir."

"Oh right. Because? . . ."

"Bollards. Can't have a lane closure without traffic control. Asking for a head-on."

"Right, so the traffic lights are for the cones. And the cones? . . ."

"Protect the traffic lights."

"? . . ."

"Well we can't have lights just stood there in the road like that without diverting the traffic, can we?"

"So the cones are for the lights and the lights are for . . . This might sound stupid, but can't you just? . . ."

"Remove it all? No sir. Traffic lights are traffic control. Bollards is protection. Without one the other's a hazard, obviously."

"You're not digging up the road, then?"

"Why would we do that?"

"Right. And you are, if you don't mind? . . ."

"Fire regs compliance. At present, the site lacks an assembly point."

"Assembly point . . . for? . . ."

"Self and Eric."

He pointed to another man reading a newspaper.

"A designated offsite site-assembly point is required to be signified—in the field there, for instance—where in the event of fire or explosion or other cause leading to site evacuation we can establish that all personnel are accounted for. Mind that cable, sir."

"You're ensuring that your own presence is compliant?"

"Exactly."

"To count two of you?"

"Indeed."

"When one of you does the counting?"

"Meaning?"

"Well, why not just not come?"

"You some kind of Marxist, sir?"

"No, I mean just don't come. Then you won't run the risk of being noncompliant."

"Then how would we know?"

"Know what?"

"If it was compliant for inspection?"

"Right . . . er . . . gotcha. . . . Thanks."

"Just serving the community, sir."

. . . .

DOES THE STORY of Norm and the bollards ring true—everything you always thought about health and safety? Funny, isn't it, how we love a story if it confirms that we're right?

We're good at filtering evidence, as we saw in Chapter 10, on big risks. The point here is that we filter it in, as well as filter it out. These stories usually have no statistical power whatsoever. But there's nothing so imperious as the smug smile of confirmation, when we discover further evidence of what we knew all along. Perhaps we need a new heuristic to counter this confirmation bias: the satisfaction heuristic, in which the more we like the evidence, the more we should distrust it. Otherwise, we'll sometimes believe stories that are absurd—provided they fit our preconceptions.

Like the story that throwing sweets into the audience at pantomimes is banned in the United Kingdom—to take one daft example. Except that it's not.[1] Rumors otherwise are just a popular myth that Britain's Health and Safety Executive (HSE) tries hard to dispel, but people who dislike what they call the "nanny state" are too willing to believe. Conker fights are okay, too, despite the story that gained notoriety in 2004 about the danger that conkers—known in America as horse chestnuts—presented to children's eyes if they shattered: "Realistically the risk from playing conkers is incredibly low and just not worth bothering about," says the HSE. But the myths persist. Loony laws are too good a story.

So for all its efforts, the HSE's reputation is probably sealed by the stories about it. "Elf and safety" is a curse, a joke. In the league of comic villains, it beats the mother-in-law. For Judith Hackitt, chair of the HSE, it's "one of the most well-worn and dispiriting phrases in the English language."[2]

And now and then, in the name of health and safety, strange things do get said, which hardly help the myth-busting. Take John Adams, who was told the windows in his block needed coating with plastic, lest people walking below during a storm, as they do, obviously, at just the spot and moment a window blew out, copped a shower of shards.[3]

Official statistics record two deaths that year from glass injury. They don't say whether these were caused by windows shattering in a storm. It seems unlikely. When Adams asked if there had been a risk assessment, he was told "it could happen," which we suppose is an assessment of sorts. Meanwhile, the lift in his seven-floor block was out of order for long stretches. There were 634 deaths on stairs that year. Norm's story of the bollards is in a long, absurdist tradition with just enough evidence to make it almost plausible. In popular demonology, Health-and-Safety-Man might as well be a villain on *Doctor Who*.

Psychologists talk of a cognitive bias for zero-risk, a preference for the certain elimination of a small risk over the probable reduction of a greater one. After all, what is safety but the absence of danger? So if you believe in a safer world, picking on risks you think you can cut out altogether may seem to make a kind of sense, almost whatever it takes.

Hearing all this, one might think health and safety was the preserve of zealots. Mind you, it can be equally tempting to see the other side—the critics of health and safety—as equally zealous in their opposition, "elf-an-safety-bureaucracy-gone-mad" ranters who fancy themselves freedom fighters.

Any chance of moderation? Because, needless to say, there's plenty of room in the middle of this argument. After all, what would you do in the John Adams case—fit plastic to the windows, mend the lift, both, or neither? We'd mend the lift, though not only for safety. We guess most people would do the same. Funnily enough, the Health and Safety Executive itself is often on that same middle ground, while it is others who whip up wild health and safety fears to stop things they don't like (see Chapter 9, on drugs, for the anthropologist Mary Douglas's ideas of risk as a form of social control).

Not all health and safety is a joke. Far from conkers and pantomimes, it is more concerned with asbestos, falls, and dangers from electricity, for example. Nor are those who campaign about it necessarily fighting over trivial, residual risks. *Hazards* magazine—another clue in the name—argued in 2012 that the HSE was ignoring evidence of an occupational cancer epidemic that kills 15,000 people a year in the United Kingdom. The magazine is free with phrases such as "corporate killers."[4]

So although irked by its extremes and absurdities, Norm has reason to be grateful to health and safety, especially for a long-term shift in attitudes toward workplace accidents. Even in 1974, when the HSE was formed, 651 employees were recorded as killed at work, representing an average risk of 29 MicroMorts per year. By 2010 this number had fallen to 120, equal to 5 MMs per year, an 82 percent decrease.[5]

Self-employment is more dangerous. Fifty-one self-employed people were killed in 2010, equivalent to 12 MMs per year, more than twice the risk for employees. Injuries have also fallen dramatically: days off from work due to injury and illness are down by about a third in the past 10 years alone.[6]

Britain also fares well in comparison with other EU countries: excluding road transportation deaths, workers in Britain were, on average, exposed to 10 MMs per year, compared with 17 in France, 19 in Germany, 26 in Spain, 35 in Poland, and 84 in Romania, according to Eurostat.[7] Tempting though it is to say that this is because the riskiest job anyone has in modern Britain is standing in a shop or sitting at a computer, Britain manufactures about as much as France (though much less than Germany).

In the United States, the Bureau for Labor Statistics provides an extraordinary glimpse of the fate of 130 million workers in 2012. It records that 4,383 were killed, representing a rate of 32 MMs per worker per year, but with a huge gender split, with an average of 6 MMs per year for women compared with 55 for men. The most common cause was highway accidents, which are excluded from the European figures. Without them, the US rate is around 24 MMs per year, about the same as Spain.[8]

The second most common cause of death in the workplace in the United States, larger even than falls, is "assault and violent acts," which make up 17 percent of all workplace fatalities. This included 463 homicides (down from 860 in 1997). So each year US workers face, on average, a risk of around 4 MMs of being murdered at work, not much less than the average risk to UK workers from all causes. Perhaps they need body armor rather than safety helmets. These homicides tend to occur in rather predictable occupations, with the risk being high for police officers (40 in 2012), workers in grocery stores (38), taxi drivers (32), bar employees (22), gas station attendants (21), and security guards (14). More surprising, perhaps, are the 7 homicides recorded in beauty salons, possibly by disgruntled customers.[9]

As we saw above, women have a much lower risk than men of being killed at work, but a greater proportion of those fatalities are homicides (29 percent versus 9 percent): essentially, female work in the service industry protects them from industrial accidents, but exposes them to random violence.

Reliable statistics about the wider world are harder to find. For example, India reported 222 fatal accidents at work in 2005, which was rather an understatement, according to the International Labour Organization (ILO), which reckoned the true number nearer 40,000.[10]

Among 2 billion workers worldwide in 2008, the ILO estimated 317 million injuries requiring more than four days' absence, and 320,000 people killed at work. Only 22,000 of these deaths were reported through official channels. The rest are a "guesstimate," giving an average of 160 MMs per year per worker, compared with 6 in the United Kingdom and 32 in the United States.[11]

All these figures are averages. They include armies of workers toiling over computers, which might cause stress, repetitive strain, and back injury but are seldom fatal, and which tend to pull the average mortality risk down. Other occupations pull it up. The history of coal mining, for example, is spotted with disaster, such as the 1,099 miners killed at Courrières in France in 1906, or the gas explosion that killed 439 in Senghenydd, Wales, in 1913. Behind these terrible events there was a steady stream of fatalities in smaller accidents, with that frightening regularity of the apparently unique and unpredictable.

Records of mining accidents began in the United Kingdom in 1850, and since then more than 100,000 coal miners have been killed and hundreds of thousands injured or suffered illness. The years 1910 and 1911 were two of the most dramatic, with violent confrontation between miners and mine-owners. Following strikes and rioting in South Wales, Winston Churchill sent in the troops. There were 1,308 deaths among 1.1 million miners in 1911—that is 1,190 MMs per year, or around 5 per shift, as if every miner on every shift went skydiving every other day.[12]

The Coal Mines Act of 1911 set up rescue stations in the United Kingdom and was intended to improve safety. Even so, around 1,000 miners were killed year after year, and in 1938 the tally was still

858 deaths, or 1,100 MMs a year. Following nationalization in 1947, mine safety steadily improved, until by 1961 there were "only" 235 deaths in accidents, a risk, after taking account of the declining workforce, of 400 MMs a year, or around 2 per shift. This improved further, but even though there are now only 6,000 coal miners in the country, a spate of accidents in recent years has brought a return to the fatality rates of the 1960s: in 2005 to 2010, the fatality rate was reported to be the equivalent of 430 MMs per year. Privatized mines were accused of cutting corners.[13]

Coal-mining accidents elsewhere are repeatedly in the news. Forty-eight miners were killed in Pakistan in 2011. In China economic growth is driven by coal, but it is deeply buried at an average depth of 400 meters (1,300 feet). Even according to official figures there is a high death toll: around 250,000 since 1949, around 7,000 in 2002, and 2,600 in 2009. Assuming 4 million miners underground, that's around 650 MMs a year (official figures), which was about the risk in Britain in 1950.[14]

Unofficial estimates put the current total nearer 20,000, or 5,000 MMs a year. This may be unsurpassed as a modern occupational risk. It is worse than Britain 150 years ago. Vast numbers of small mines are run by corrupt local officials under minimal central control. The Asian Development Bank has commissioned worthy reports, but attempts to impose safety regulations or closures often lead only to the growth of illegal mines.[15]

The highest-risk occupation in the United Kingdom today is commercial fishing. A recent study recorded 160 deaths between 1996 and 2005, which works out as 1,020 MMs per year per fisherman. Fourteen fishermen died in coastal shipwrecks, 59 drowned when their boat sank or capsized, mainly because it was unstable, overloaded, or unseaworthy. About half the fatalities are solitary fishermen, not helped by a culture of shunning personal flotation devices. Perhaps unique among industries, these risks have not declined since World War II, although before that they were even higher: 4,600 MMs per year in 1935–1938, close to the unofficial risk for Chinese miners today.[16]

In the United States, logging topped the occupation-risk tables in 2012, with 62 deaths, equivalent to 1,280 MMs a year, while fatalities in US commercial fishing came in second, at 1,170 MMs a year—almost exactly the same as the UK rates, and very similar to the 1,000 to 1,160 MMs per year for Danish, French, and Swedish fishing fleets. The risks are even higher in New Zealand fishing fleets (2,600). The next most hazardous occupations in 2012 were considerably lower for the United States: airline crews (534 MMs) and roofers (405).

Of course, it is not just workers who can suffer from industrial accidents—bystanders are caught up, too. A famous example was the great London beer flood of October 17, 1814, when huge vats of porter burst in the Meux brewery on the corner of Oxford Street and Tottenham Court Road.[17] More than a million liters of strong beer—enough to fill 20,000 barrels—broke through brick walls, demolished two houses, seriously damaged the Tavistock Arms pub, and poured into cellars occupied by the local poor. Nine people drowned, and the fourteen-year-old barmaid Eleanor Cooper was crushed by rubble in the Tavistock Arms. Others, of course, rushed to the scene to fill up pans and kettles. The story that one fell victim to acute alcohol poisoning could be an urban myth. In the later court case the disaster was ruled to be an act of God, and the brewery escaped responsibility.

Just over a century later, this bizarre event was eclipsed by the Boston molasses disaster of January 15, 1919. This event was like a B-grade movie. A huge tank of 8 million liters of molasses burst—equal to around three Olympic swimming pools—and a black, sweet, sticky tsunami, 3 meters high and traveling at 35 miles per hour, swept out, demolishing buildings, damaging the overhead railway, drowning numerous horses, killing 21 people, and injuring 150 more. The United States Industrial Alcohol Company was found responsible and paid compensation.[18]

If there's black comedy in these stories of absurd accidents, events in Bhopal in 1984 are of a different order. Thirty metric tons (about 33 US short tons) of methyl isocyanate (MIC) leaked from a tank belonging to a subsidiary of the US Union Carbide Corporation. MIC is a gas, but heavier than air. There was no escape from nearby shanty towns. Three thousand people died immediately. Recent estimates suggest a final death toll of 25,000, as well as half a million injuries, many permanently disabling. The court cases continue.[19]

The victims of Bhopal illustrate the distressing long-term effects of industrial accidents and exposures. The ILO estimates that 2 million people worldwide died from diseases caused by their jobs in 2008. In the United Kingdom, the HSE estimates that in 2009 there were around 8,000 cancer deaths due to people's previous occupations, half of them from asbestos.[20] In the United States, deaths associated with pneumoconiosis, caused mainly by mining, have been falling—from 4,151 in 1980 to 2,028 in 2010. Deaths from mesothelioma (cancer caused by asbestos) are still steadily rising and have now overtaken pneumoconiosis, increasing from 699 in 1980 to 2,744 in 2010.[21]

The risk of fatal illness in later life is a chronic risk rather than an acute one. When exposure to asbestos kills, it usually does so long after the event. It would be useful to know the chronic risk faced by asbestos workers or miners in terms of lost MicroLives per day of employment. But something strange happens in this calculation. The long-term mortality of asbestos workers turns out to be only 15 percent higher than that of the general population, roughly corresponding to the loss of around 2 MLs per day, no worse than a few cigarettes. Even more paradoxical is the finding that the mortality of 25,000 British coal miners was 13 percent lower, on average, than for men living in the same regions.[22]

The explanation for these findings is known as the "healthy worker effect." Men would not be down the pit unless they were fit and healthy, and so, compared with average people, miners can appear to have better survival in spite of their exposure to coal-dust, because they are drawn from a healthier population. This makes it hard for epidemiologists to calculate the harm of poor working conditions except in general terms—or by counting bodies.

We've seen how some occupations are still risky, although not at Victorian levels. But how high must the risk be before the HSE decides that something should be done? The HSE's philosophy is based on what is known as the "Tolerability of Risk" framework, and can be nicely converted to MicroMorts.[23]

Potential hazards are thought of on a spectrum of increasing risk, divided into three loosely defined categories. At the top are "unacceptable" risks: whatever the benefits of the activity, something must be done to protect either the workers or the public, or both. At the bottom are "broadly acceptable" risks: not quite zero but considered insignificant, the sort of thing we would regard as normal in our daily lives.

In between these extremes lie the "tolerable" risks: those we might be prepared to put up with if there is sufficient benefit, such as providing valuable employment, involving personal convenience, or keeping the infrastructure of society going: somebody has to do the dirty work. The HSE makes clear that risks should be considered tolerable only if they are carefully weighed using the best evidence, are periodically reviewed, and are kept "as low as reasonably practicable," a set of criteria known as ALARP. (The HSE has other great mnemonics that are not so easy to pronounce: duties to reduce risk may be SFAIRP—"so far as is reasonably practicable.") But how does anyone decide what is unacceptable, broadly acceptable, or tolerable? The HSE suggests some rough rules of thumb.

First, it states that an occupational risk might be considered "unacceptable" if the chance of a worker being killed is greater than 1 in 1,000 per year, or 1,000 MicroMorts per year. On this metric, the risks faced by pre-nationalization British miners would now be considered "unacceptable," and current commercial fishing is around this level. The HSE excludes "exceptional groups": presumably serving in a war zone counts as exceptional—for example when the average risk faced by 9,000 servicemen and women in Afghanistan reached 47 MMs a day, or around 17,000 a year in late 2009.[24]

For members of the public, rather than employed workers, the HSE considers a 1 in 10,000 a year risk—that's 100 MMs a year—as generally unacceptable. At the other extreme, risks are considered broadly acceptable if they are less than one in a million per year—1 MicroMort—the current estimate of the average risk of being killed by an asteroid.

Even such a minimal risk, if applied to the whole population of the United Kingdom, would mean around 50 deaths a year. This brings another curiosity to the forefront. Imagine the headlines if 50 deaths happened at once. As we've seen when discussing railway disasters in Chapter 15, cold MicroMort calculations are easily trumped by "societal concerns" about disasters with multiple fatalities, hazards that affect vulnerable groups such as young children, or risks imposed on people just because of where they live.

The HSE offers separate guidance that reflects this: the risk of an accident involving 50 deaths should be less than 1 in 5,000 per year. Spread across a population of more than 10,000, it is less than 1 MicroMort a year each. From an individual perspective this might be considered "acceptable," but because we don't like disasters (although people clearly love reading about them in newspapers), huge amounts of money are spent to make tiny risks even tinier.*

---

* Such explicit consideration of "tolerable" fatality risk is not part of the official discussion in the United States, where the explicit pricing of a year of life, as used by the UK National Institute of Health and Clinical Excellence, is not acceptable as part of health-care policy. This is no place to explore the possible reasons for these cultural distinctions.

19

RADIATION

Prudence scanned the shelves of vitamins, tonics, and remedies at the drugstore. Pansy swung from her mother's arm. Above them in a bundle tied by their feet, soft toys wore the label "I'm a sloth."

"What are those, mommy?"

Prudence looked up.

"Sloths."

"Can I have a sloth?"

"No."

"What do they do, mummy?"

"Nothing."

"Why not?"

"They just sit there."

"Please mommy?"

"Stuffed."

"Pleeease?"

"Doing nothing."

"Can I?"

"Like your father."

Prudence disapproved of soft toys. Think of the dust mites and plastic eyes. She disapproved of her husband's inertia, too. His cough, for instance.

She took from the shelf a small bottle of thirty-seven life-giving essential vitamins, minerals, and nutrients for health and vitality, including active antioxidant extracts of Korean ginseng, and tossed it into the basket with two others of high-strength organic sunscreen.

"It's a cough," he said that evening, as they sat watching the news about an accident at a nuclear power station, while Prudence ate her ten o'clock banana.

"Can't a man have a cough and it just be a cough?" he said, then added, mysteriously: "Fires sometimes go out."

But the story on the news was of a fire that didn't go out. "Meltdown," they said, and they talked of "criticality excursions," which sounded like a fatal bus ride. Why he agreed with it, in his boys-and-lasers way, she'd never know. The geeks were clueless, it wasn't natural, and she'd rather live in a mud hut than be "saved" by technology like that. Creepy stuff.

"So you'd leave a fire burning to see if it went out?" she said.

"What? No. I mean . . . "

"What if? *What if*? Think if we lost you."

"Hmmm," he said.

He had turned down her offer of an all-body scan for his birthday. The ultimate in preventive health screening, peace of mind, a 3D computerized journey through your own body with a free colonoscopy and take-home DVD to share with friends.

"Look at that," she said, reading the testimonials. "Just an ache and it was kidney cancer."

There was a picture of tanned and attractive, smiley doctor and technician types in white coats gazing at computers—and in the background a man's bare legs in black ankle socks sticking from a tube.

"Tumors, cysts, hemorrhage, blockages, heart, bones—your back, for instance—infection. It's quite comprehensive."

"They put you in a machine," he said. Cough.

"They're professionals."

"Doctors? Ha!" Cough.

"Well, rather one machine than a lifetime's worry."

"I wasn't worried . . . " Cough.

. . . .

PRUDENCE'S HUSBAND would be irradiated by an all-body scan—for the sake of his health. Good idea, she thinks. Nuclear power is a bad idea—because of the radiation, she thinks. Some contradiction, surely?

He, on the other hand, admires the nuclear physicists' nerdy mastery of nature for the common good, but thinks that doctors who use radiation technologies are quacks. Is he any more consistent?

Hearing and telling stories about risk, people like to think they react in a consistent and justifiable way. Do they?

Since exposure to radiation can be measured in standard units, we know how one exposure compares with another. We know roughly how lethal they are, on average, from the good (X-rays) to the bad (radon gas in the home) to a mixture of both (a suntan). This makes radiation a handy test case, in that people's differing sense of risk about the same level of exposure from different sources is a clue to what personal risk is often about.

What it's often about in the case of radiation is known—no mincing of words here—as "dread." For Prudence, at one extreme, the threat of being fried or contaminated by an accident at a nuclear power plant ticks all the fear-factor boxes that tend to be invoked in definitions of dread risk. For example, it is an invisible hazard, mysterious, ill understood. It seems unnatural. It is associated with those particular evils cancer and birth defects, provoking concern for future generations, with a vague feeling of catastrophic potential. Prudence feels she can't control or avoid it, which makes it worse—it is involuntary.* And it is associated with visceral reaction, not calculation.

---

* The media have a field day, even with the very word "radiation," although heat, light, and radio waves are forms of radiation, too. So it's important to distinguish between these "non-ionizing" types, which nobody much cares about (except those who feel they are being harmed by cellphone towers), and the "ionizing" sort, which potentially has sufficient energy to cause changes to atoms and which concerns us here. Ionizing radiation can damage cells, which is why radiotherapy is used against cancer, although it is still not clear whether very low doses are harmful. Very high doses can cause acute radiation sickness, while the late effects of radiation exposure can increase the risk of cancers by causing cell mutation.

We've touched on many of these in previous chapters, but separately. "Dread" can arise from any one of them, and it positively soars when the fear factors mix or multiply. For example, falling head first from a playground climbing frame would be a relatively low-dread event for most people, whereas radiation poisoning would usually be a high-dread event, almost regardless of the probabilities of fatality for any individual.

Dread is often described as "disproportionate," "excessive," or "irrational," words that don't try to hide the disdain of radiation professionals. If only people weren't so ignorant of the technicalities. And it is true that people react differently to similar radiation risks, no question. So it looks as if a case could be made that dread can make them overly emotional and distort their perceptions. But are these dread reactions silly, and even dangerous and deceiving in themselves, or does the notion of dread risk catch something the numbers miss?

When radiation is used to help diagnose or treat medical problems, most people are like Prudence—relatively untroubled. The history of its medical use has been sublimely blasé. From the first use of X-rays in the 1890s, the magical ability to see inside the body made the harm they could cause seem negligible. Never mind that Thomas Edison's assistant, Clarence Dally, and numerous radiologists died early. After radium was discovered by the Curies, radioactivity was thought of as a source of health and healing. In 1909, Dr. E. Skillman Bailey reported to the Southern Homeopathic Medical Association in New Orleans on his investigations of Radithor, which enabled him to "photograph objects through 6 inches of wood." He used it for medical treatments, although he was no doctor and was said to exhibit "visible signs of being unstrung." He even passed a sample around the audience. Nevertheless, Radithor—"Radioactive water, a cure for the living dead"—was a success and sold hundreds of thousands of bottles until the company was forced to close in 1932, owing to bad publicity following the death of the steel millionaire and playboy Eben Byers from radiation poisoning. Byers had consumed more than 1,400 bottles. The *Wall Street Journal* headline was: "The Radium Water Worked Fine Until His Jaw Came Off."

There were no radiation safety standards until the 1920s. They've grown increasingly stringent ever since: X-ray "Pedascopes" for fitting children's shoes were banned, as was the "treatment" with X-rays of children with ringworm, and mental patients with radium. Nevertheless, X-ray and computerized tomography (CT) scans are still carried out in huge numbers.

People also seem rather unconcerned about natural background radiation arising from sources such as cosmic rays and radon, a radioactive gas from uranium in rocks such as granite. Radon, which can collect in ill-ventilated rooms, is estimated to cause 1,100 preventable lung cancer deaths in the United Kingdom each year. But these exposures do not arouse the strong feelings associated with nuclear power.

John Adams cites cellphones as another example of apparent inconsistency that picks up on one aspect of dread: choice, writing: "The risk associated with the handsets is either non-existent or very small. The risk associated with the base stations, measured by radiation dose, unless one is up the mast with an ear to the transmitter, is orders of magnitude less. Yet all around the world billions of people are queuing up to take the voluntary handset risk, and almost all the opposition is focused on the base stations, which are seen by objectors as impositions."[1]

Since the phone is useless without the transmitter, that appears odd. Except that you can't choose where the transmitter goes, and you can't help wondering why not. What are they up to, these transmitter people? Which shows how these apparently inconsistent attitudes might make sense when seen as an expression of feelings about freedom of choice. We say: "It's risky." What we might partly mean in these cases is: "No one asked me."

In a now classic paper from 1989 on the differences between lay and expert opinion about risk, Paul Slovic said that "dread" expressed a subtlety that the experts sometimes overlooked. "There is wisdom as well as error in public attitudes and perceptions. Lay people sometimes lack certain information about hazards. However, their basic conceptualization of risk is much richer than many of the experts and reflects legitimate concerns that are typically omitted from expert risk analysis."[2]

Although we can acknowledge these concerns, it is still valuable to get a handle on the magnitude of the hazard. So what are the exposures? The units are confusing. It's easiest to focus on the Sievert, which is a measure of the biological effect of being exposed to radiation. One Sievert—or 1 Sv—can cause radiation sickness, which might mean hair loss, bloody vomit and stools, weakness, dizziness, headache, fever, skin redness, itching or blistering, infections, poor wound healing, and low blood pressure. Not nice.

One Sievert can be broken down into 1,000 milli-Sieverts (1 milli-Sievert—or 1 mSv—is the US Environmental Protection Agency limit for the annual exposure of a member of the public) and 1 million micro-Sieverts. A tenth of 1 micro-Sievert is about the dose from eating a large banana—Prudence's evening snack—or going through a whole-body scanner at an airport.[3]

The radiation dose chart, Figure 23, is a fun but imperfect way of comparing radiation from the mass of different sources by converting Sieverts into a banana equivalent dose, or BED (imperfect mainly because the radiation in bananas comes from potassium, and the body can regulate this). The virtue of this whimsical unit is that it shows the enormous range of exposures on one scale. Five hundred million bananas—depending, of course, on the size of the banana—is about 50 Sieverts, the equivalent of 10 minutes next to the Chernobyl power station during meltdown. This helps make the point that many hazards are not a hazard at all if the dose is low enough. After all, who's worried about one banana?

This chart is a minor publishing scandal, by the way, and may get us into trouble. Most advice on risk communication is to avoid mixing voluntary with involuntary risks. But this table mixes these risks freely, together with a variety of ways of comparing them. Our excuse is that if emotions make people immune to data, then maybe one way to even up the contest is to make the data emotionally shocking, perhaps by mixing up surprising comparisons. Bananas, CT scans, and Chernobyl? You bet. Just like horse-riding compared with Ecstasy, such comparisons are outrageous, but fascinating, and

Exposure	milli-Sieverts	Bananas	Equivalent distance from epicenter of Hiroshima explosion	Average loss in life-expectancy	Cigarettes
Ten minutes next to Chernobyl reactor core after explosion and meltdown	50,000	500 million	100 m	50 years	200,000
Dose of radiation that would kill about half of those receiving it in a month	5,000	50 million	700 m	5 years	20,000
Acute radiation effects, including nausea and a reduction in white blood cell count	1,000	10 million	1.1 km	1 year	4,000
Annual exposure limit for nuclear industry employees	20	200,000	2.2 km	7 days	700
Effective dose in worst-affected areas around Fukushima nuclear plant	10–50	100,000–500,000	1.9–2.4 km	3–30 days	300–1500
Whole-body CT scan	10	100,000	2.4 km	3 days	300
Average annual radon dose to people in Cornwall	8	80,000	2.4 km	3 days	300
Chest CT scan	7	70,000	2.5 km	3 days	300
Effective dose in Fukushima prefecture	1–10	10,000–100,000	2.4–3 km	10 hours–3 days	30–300
A year of normal background dose, 85 percent of which is from natural sources	2.7	27,000	2.7 km	1 day	100
Mammogram	0.4	4,000	3.2 km	4 hours	16
Nuclear power station worker average annual occupational exposure	0.18	1,800	3.5 km	2 hours	8
Approximate total dose received at Fukushima Town Hall in the two weeks following accident	0.1	1,000	3.6 km	1 hour	4

FIGURE 23. Approximate radiation exposure from different sources, converted into banana-equivalents, equivalent distance from epicenter of Hiroshima explosion, and average loss in life-expectancy. Note: 1,000 micro Sieverts (10,000 bananas) = 1 milli-Sievert; 1,000 milli-Sieverts = 1 Sievert. *(continues)*

Exposure	milli-Sieverts	Bananas	Equivalent distance from epicenter of Hiroshima explosion	Average loss in life-expectancy	Cigarettes
Flight from London to New York	0.07	700	3.7 km	37 mins	2
Chest X-ray	0.02	200	4.1 km	11 mins	1
135g bag of brazil nuts	0.005	50	4.4 km	3 mins	0.2
Dental X-ray	0.005	50	4.4 km	3 mins	0.2
Eating a banana (or going through airport scanner)	0.0001	1	5.5 km	3 secs	puff
Sleeping with some-one	0.00005	0.5	5.7 km	1 sec	Small puff

FIGURE 23. (continued)

in our view genuinely help to put risks in proportion, partly by forcing us to think about the basis for our emotional reaction to different hazards. If perceptions are what often determine our sense of risk, then surprising perspectives have a part to play.

Most of our knowledge about the harmful effects of radiation comes from detailed studies of the victims of the bombs dropped on Hiroshima and Nagasaki in 1945. Around 200,000 people died immediately or within a few months, but 87,000 survivors have been followed up for their whole lives. By 1992, more than 40,000 of them had died, although it's estimated that only 690 of those deaths were due to the radiation. Bomb victims had been exposed to radiation roughly according to the dosage shown in the table: the dose declines rapidly, halving every 200 meters (650 feet), and so, according to a US National Academy of Sciences report, someone around 2.5 kilometers (1.5 miles) from the epicenter received around the same dose as from a modern CT scan.[4] What would Prudence say?

The number of victims of the Chernobyl accident is more controversial: a United Nations report says that acute radiation sickness has killed 28 people, and that 6,000 children developed thyroid cancer by drinking contaminated milk, an easily preventable event. Of these, 15 had died by 2005, but the report adds that "to date, there has been no persuasive evidence of any other health effect in the general population that can be attributed to radiation exposure."[5]

Others claim the true figures are far higher. It all depends on what you believe about the effects of low levels of radiation. When experts say there is no evidence of extra harm, this is hardly surprising, as even if there had been an increase in cancers downwind of Chernobyl, they would be impossible to detect, given the vast numbers that occur anyway. So any estimate would have to be based on a theoretical model of harm.

Using these models, it is estimated that, on average, roughly one year of life is lost per Sievert exposure.[6] When setting regulations, these effects are extrapolated down to much lower doses for which there is no direct evidence of harm; this is known as the Linear No Threshold (LNT) hypothesis. If we assume LNT, then 1 mSv is 1/1000th of a year lost, which is 9 hours of life, or 18 MicroLives. So, as a mammogram is 0.4 mSv, this is equivalent to 8 MLs, or around 16 cigarettes. Both average loss in life-expectancy and cigarette equivalent are shown in the table.

The LNT hypothesis is controversial: the people working out the effects of Chernobyl did not use it, and some claim that low doses do not have a proportional effect, since the body has time to heal itself. But if we accept the LNT idea, then we can get some remarkable conclusions about the "nice" use of radiation to help sick people.

For example, a CT scan of 10 mSv comes in at 180 MLs, or around 360 cigarettes. This may not seem too large for an individual getting a diagnosis, but over large numbers of people it adds up: the US National Cancer Institute estimated that the 75 million CT scans in the United States in 2007 alone will eventually cause 29,000 cancers.[7]

None of this was discussed when the apocalyptic visions of destruction brought about by the Japanese earthquake and subsequent tsunami in March 2011 were largely replaced in the media by reports of the struggle to control radiation from the stricken Fukushima nuclear plant. This ticked all the boxes for dread:

invisible, uncontrollable, associated with cancer and birth defects. Add an untrusted power company into the mix and the psychological consequences are predictable. The EU Energy Commissioner, Günther Oettinger, told the EU Parliament that "there is talk of an apocalypse and I think the word is particularly well chosen."[8] Was it? From what perspective?

So should Prudence encourage her husband to have his CT scan? Would Prudence willingly stand a mile and a half from the Hiroshima bomb? Same dose, maybe. Same risk, perhaps. But is she really irrational to say yes to one and no to the other?

When people look at comparable risks of exposure and regard one with horror but the other with approval, it might reveal that they are thinking with their gut. But might it also reveal the limitations of probability as a measure of what danger is really about? The advantage of trying to look beyond words like "irrationality" or "inconsistency" is that then we stand a better chance of finding out why people really dislike things. To reply, "well obviously, because of the danger," is a convenient argument, but often not even close to the whole human truth.

20

SPACE

When the body of the stowaway hit the street, it sounded like a slamming door. He was frozen, which, all things considered, was probably for the best. But Prudence didn't think so.

"What if it had hit the sunroom?" she said.

"He, not it," Norm said. "Anyway, it didn't."

"Norm, we sit there for breakfast. Pansy was *there, on her own*, on *that* chair, when . . . my God . . . "

"Pru, the chance . . . "

"Chance?! What were the chances of a dead body falling?* I mean, one [she raised her finger], out of the air; two [another finger], to the ground, all the way; three, in Basingstoke; four, in the first place? Tell me that. And what happened? It happened. Well, if that can happen . . . You see. Every morning, we sit in that sunroom having breakfast. Next thing you know there's a stiff Algerian in your Weetabix. It only takes one, you know."

But a few days later it was Norm's turn to look up anxiously.

"You see, asteroids could be it," he said to Prudence, turning his teacup. It was a lottery, entirely, the chance of the big hit by one among millions of possible hits in an infinity of cosmic debris.

---

* An Angolan man fell from an airplane onto a residential street in Mortlake, London, in September 2012. Jon Kelly, "How Often Do Plane Stowaways Fall from the Sky?" BBC, September 13, 2012, www.bbc.co.uk/news/magazine-19562101.

He turned the cup again. It wasn't fear of death that bothered him; more like working out the odds. One day something would fall from the sky, and the chance was both certain and invisible. It made him cross.

"Oh, I know," said Pru. "I'm the same with pigeons."

"It's not calculable, the probability, not sensibly."

"Well it's bloody amazing, the aim."

"And the damage . . . the carnage, potentially."

"The cleaning, that's for sure."

"Think of London."

"Quite."

"I'm coming round to some form of general-purpose defense."

"An umbrella?"

"Technically speaking, an umbrella, yes."

"Always have one."

"Despite the cost."

"Trivial by comparison, surely?"

"Absolutely."

Later, as night fell, he looked up. For a long time he stared, a man alone under the stars, looking for answers. Where was it, the one with his number on it, out in the vastness, the blackness, the unknown? Somewhere, for sure, probably. Silent and unseen, hurtling toward earth's small plot in this . . . this . . . fathomless . . . this universe, this little O. The path, the uncluttered, no cluttered, the cluttered path through space of the asteroid hurled by a God of limitless thunder that one day would end all human ends was one possible, improbable path . . . possibility among countless others possibly . . . possible path probabilities, probably. One lump, grain—because it's all relative—in a boundless storm, yes, and the two trajectories of earth and rock predestined by infinite lottery to meet by a cosmic whisker. It was the ultimate doom, reckoning, fate, and embodiment of all fear.*

---

*Contemplating human fate while staring at the cosmos has a funny effect on people's prose. See Carl Sagan, who manages proper lyricism in *Pale Blue Dot: A Vision of the Human Future in Space* (New York: Random House, 1994).

"So," said Kelvin in the pub later, "what's it mean for house prices?"

"?"

"Practicalities, Norm, come on."

Email: *Prudence to Norm*
Subject: *Apocalypse*
Norm, dearest,

The asteroid was on the news, as you said. Too soon to tell, they said, but could be a chance. Wondered if you knew any more gossip among the star-gazing fraternity, such as whether the devastation would be global, as we were thinking of that holiday home in Portugal.

"I was just wondering, love, about the asteroid," said Mrs. N a few days afterward, "whether it was worth remortgaging—I mean, if we never had to repay? . . ."

. . . .

NOT LONG AGO, Norm discovered he was l'*homme moyen* and hit the peak of hope and self-belief. Now he faces an existential crisis. Why? Because with the danger from asteroids he confronts one of the most absurd averages ever in the ultimate life-or-death calculation. That makes it not just a distant threat to survival but an immediate threat to everything Norm stands for. We'll come to this strange average and the calculation behind it in a minute.

First, who has the more reasonable fear: Norm or Prudence? Heavenly bodies falling from space or human bodies from planes? And why?

Prudence has one advantage: familiarity. She can imagine bodies, planes, and pigeons easier than apocalypse—for obvious reasons of experience. Easier, too, perhaps, to get your head around cosmic doom by thinking in terms of house prices.

According to news reports, the Heathrow flight path has seen a few falling bodies over the years, the tragic results of desperation and the freezing, oxygen-starved atmosphere above 30,000 feet. In 2001, the body of Mohammed Ayaz, a twenty-one-year-old Pakistani, was found in the parking lot of a hardware store in Richmond, London. Four years earlier, a stowaway fell from the undercarriage of a plane onto a nearby gasworks. No one was hit.

Nobody was at home at the house of the perfectly named Comette family in Paris, either, in the summer of 2011, when an egg-sized meteorite seared into the roof.[1] The rock, blackened as it passed through the Earth's atmosphere, smashed a roof-tile and buried itself in the insulation. It wasn't until Martine Comette noticed the rain coming in and called someone to fix it that she discovered the cause. The rock was thought to be about 4 billion years old, from a belt of asteroids between Mars and Jupiter. When Martine's son Hugo took it to school in a piece of kitchen roll, his friend said it looked like a lump of concrete.

A few months later, in September 2011, a NASA satellite fell somewhere off the West Coast of America—prompting concern about how likely it was to land on a human head. At the same time, Lars von Trier released his film *Melancholia*, in which "two sisters find their already strained relationship challenged as a mysterious new planet threatens to collide with the Earth," all to music from Wagner's *Tristan and Isolde*.

All of which gives the impression that there is a lot of heavenly debris about. So what are the risks that a solid object will appear from space and land on your head?

For heavenly bodies the calculation is tricky, partly because of the chance of a truly big hit. And this is what Norm finds so disturbing. An insurer dealing with a car-on-car collision has abundant direct historical data to help calculate the risks. Astronomers have little. Instead, they devise equations relating to the size of an asteroid, how many of them are out there, how often they might hit the earth, and what the explosive force would be. These estimates are continually revised and subject to esoteric dispute. The average bottom-line risk that they produce turns out to be truly absurd. We'll come to it in a moment.

There are two main considerations in calculating potential damage. First, the size of the object; second, where it strikes. If a tree falls in a forest and there's nobody there to hear it, does it make a noise? And if an asteroid strikes the atmosphere at a speed of 15 kilometers* a second and explodes 10 kilometers above a Siberian forest, where it flattens an area of trees measuring 40 by 40 kilometers but scarcely anyone is there to see, does it matter? When this happened on June 30, 1908, in Tunguska, the few eyewitnesses willing to talk spoke of a heat so intense it felt as though their clothes were burning, even 65 kilometers away, of a sky split in half with fire, of being knocked from their feet and running in panic, thinking the end was nigh. It was dramatic, leaving 80 million scorched and flattened trees; otherwise there was some confusion, not many hurt, and no direct evidence that anyone perished.

Had the meteorite landed 4 hours and 47 minutes later, it would have hit St. Petersburg, or so it has been calculated. One estimate is that such an airburst over New York today would cost $1.19 trillion to insurers in property damage, not to mention roughly 3.2 million fatalities and 3.76 million injuries.[2]

So where an object lands is of at least equal significance to what it is. There is a lot of stuff flying around up there: asteroids made of rock, comets made of ice, and frozen gases. At the smallest scale—according to the riveting US National Research Council (NRC) document *Defending Planet Earth: Near-Earth Object Surveys and Hazard Mitigation Strategies*—around 50 to 150 metric tons of "very small objects," mainly dust, drop to Earth every day. Simply looking into a clear night sky can reveal regular trails of rock or dust burning up in the atmosphere.[3]

Bigger—but only a little more serious—asteroids of 5 to 10 meters in diameter come by around once a year, releasing energy equivalent to around 15 kilotons of TNT when they explode in the upper atmosphere, about the same as the Hiroshima bomb. Most go unseen and unrecorded.

---

* We have used metric measurements throughout this chapter, following the usual practice in astronomy and other fields of science. Conversion calculators can be found online for readers interested in finding US equivalents.

Occasionally something gets through and leaves a visible crater or disappears harmlessly into the sea. There is no recent record of human fatality from a meteorite strike, although a few cars in the United States have been damaged over the past century, and one, the famous Peekskill meteorite car, a Chevrolet Malibu with a trunk that looks as if someone took a sledgehammer to it, has toured the world. A cow was also killed on October 15, 1972, in Valera, Venezuela, and duly eaten. Bits of the meteorite were sold to collectors.

A little bigger again, an airburst of an asteroid about 25 meters in diameter—about the volume of 60 or 70 double-decker buses—would release energy of around 1 million tons, or 1 megaton (MT), of TNT, equivalent to around 70 Hiroshima bombs. The Tunguska meteorite is thought to have been about 50 meters across, although some astronomers suggest that Tunguska-like events could be caused by objects as small as 30 meters in diameter.[4] Up to this point, even if such an asteroid does strike the earth, there is about a 70 percent chance that only water will be in the way. So we might still be lucky. Unlucky, and it could kill millions.

Still bigger asteroids begin to fall into the range described as "continental-scale events," although again, it is tricky to say what damage such an impact would cause. On land, it could be devastating. As before, there is a 70 percent chance it would strike ocean, but this time with more consequence. Models have suggested that an asteroid 400 meters in diameter could cause a tsunami 200 meters high—almost twice the height of the Statue of Liberty, ground to torch—although, of course, there is great uncertainty about whether such a wave would break on the continental shelf, whether a population could evacuate, and so on.[5]

In the seriously big league, an asteroid more than 1 kilometer across would release around 100,000 megatons of energy: equal to about 700,000 Hiroshima bombs and potentially globally catastrophic. Even bigger collisions can occur and have occurred. More than a cow perished when a lump 10 kilometers in diameter hit the Yucatán peninsula in Mexico 65 million years ago with a force of about 100 million megatons. It left a crater more than 180 kilometers in diameter and probably wiped out the dinosaurs.

That gives some sense of the potential range of damage. The next step is to discover how many of these things are out there and from this derive an estimate of the likelihood that Earth will be in their way.

Fortunately, NASA's Near-Earth Object Program is watching over us, and reported, for example, that object 2012 DA14, about 45 meters across, was due to pass within 17,000 miles of earth on Friday, February 15, 2013. By a delicious coincidence, worthy of this book, 16 hours earlier an undetected asteroid exploded above Chelyabinsk in Russia with an estimated force of 500 kilotons, injuring more than 1,000 people—this was around 20 meters across—the largest object to hit us since Tungaska, and the first known to have injured people.

Near-Earth Objects (NEOs) are those which could come within one-third of the distance to the sun—around 50 million kilometers—showing that "near" is a relative term. When Perry Como sang "Catch a Falling Star and Put It in Your Pocket" in the 1960s, there were only 60 NEOs known, but by December 2013, more than 10,400 had been found and named by the Minor Planet Center at the Smithsonian's Astrophysical Observatory. Each year about another 500 are added to the list. There are enough of these for the NRC to estimate an impact equivalent to the devastation from the Tunguska meteorite in Siberia about every 2,000 years, on average, although, if it's true that Tunguska-like events could be caused by objects as small as 30 meters across, they would occur 10 times more frequently, giving odds of a hit of this size of nearly 50:50 in anyone's lifetime.[6] An even chance that a baby born today will see a Tunguska equivalent, capable of wiping out New York City, is about as scary as asteroid statistics can be made to appear. But it is contentious: analysis following the airburst over Chelyabinsk has suggested that the risk of such impacts may be 10 times higher than previously estimated.[7]

Of the big, wipe-out-sized NEOs more than 1 kilometer across, 834 asteroids and 90 comets have been identified, and NASA's Near Object Program estimates that there are only around 70 left lurking out there that we don't know about.[8]

A little closer to home than the 50 million kilometers that qualifies as "near Earth" is 20 times the distance to the moon (a total of about 8 million kilometers), within which range any lump of rock that is found to be at least 150 meters across will earn Potentially Hazardous Asteroid (PHA) status. By December 2013, 1,442 PHAs have been found, 151 of which are in the potentially end-of-the-Earth-as-we-know-it bracket of more than 1 kilometer in diameter.

Still, these quantities are low enough to suggest that a clash with debris of that caliber is only expected every few million years. The NRC stoically points out that "while this apocalyptic possibility is extraordinarily unlikely to happen in the lifetime of anyone living now, traditional approaches to preparing for disaster would become irrelevant." Have they not heard of Bruce Willis? Mass extinctioners of the 10-kilometer size that wiped out the dinosaurs are estimated to come along every 100 million years or so, so not much need to worry there.

These are all averages, of course, seen from a perspective of thousands or millions of years, but the NASA catalog of NEOs allows us to escape just a little from largely theoretical calculations and talk instead about what might be in store for those of us living now from specific rocks, all with their own personalized names or numbers. Each NEO is classified according to its size and probability of impact: these probabilities do not reflect any randomness, since it is assumed that the asteroid is either going to hit us or not, but is simply due to ignorance of its precise trajectory.

The TORINO scale[9]—named after the conference venue in Italy at which it was adopted—expresses the appropriate level of concern for each asteroid:

Level 0 (white) says no problem
Level 1 (green) is a routine safe fly-by
Level 2 (yellow) merits attention by astronomers but not the public
Level 3 means the public should be told

And so on up to Level 10—a certain collision threatening global civilization.

At any time there are usually a few at Level 1, but Apophis, which is around 200 to 300 meters across, was given Level 2 status in 2004, when it was discovered, and temporarily upgraded to Level 4 when a 2.7 percent chance of impact in 2029 was estimated.[10] New information has shown that we're safe after all, although Apophis should be visible to the naked eye on April 13, 2029, when it passes about 30,000 kilometers from Earth. Note that April 13 that year falls on a Friday.

The current catalog indicates no serious risk from asteroids that we know about. In December 2013, the greatest danger that NASA could point to was a 1 in 2,000 chance of a collision with the 130-meter 2007-VK184 sometime in the 2050s, and that is only Torino Level 1. But most asteroids less than 500 meters across remain undiscovered, and although these are unlikely to bring about the end of the world, they can take us by surprise: 2008 TC3 was around 2 to 5 meters across and weighed some 80 metric tons when it burst above the Sudanese desert on October 7, 2008. It was the first such asteroid to be detected before impact, but there was only 19 hours' warning, and one can imagine the crisis if the predicted path had contained a big city. It exploded with a force of around 2,000 metric tons of TNT, and 10 kilograms of bits were picked up afterward. Nobody was hurt.

So what is the risk of being killed by one of these rocks from space in your lifetime? This is impossible to assess accurately, and that's where Norm comes unstuck. For although there is rapidly increasing understanding of the chance of an impact, predicting the consequences requires many assumptions (or guesses, depending on your point of view). Nevertheless, the NRC report quotes the wonderfully precise figure of 91 deaths expected per year. Of course, this is an average between almost all years with zero deaths and a few events, averaged over millennia, with massive casualties, which, once again, just goes to show the problem with averages. In fact, that 91 is roughly evenly balanced between the more common small-scale impacts and the very unlikely globally catastrophic impacts. Since there are 7 billion people on earth, this works out at 1/77th of a MicroMort per person per year—the equivalent of about a 3-mile car journey—which comes to the delightfully round number of 1 MicroMort per lifetime from asteroids. Not a lot. And a faintly ridiculous number with no practical purpose. It's not surprising Norm is troubled.

What could be done about an imminent threat? The NRC report identifies four main strategies for mitigation, emphasizing the massive uncertainty about the hazards, the technology, and how society would respond. The first, *civil defense*, involves standard disaster management and is suitable for small events, or anything without much warning. If, for example, it was found that Apophis did have a high probability of impact in 2029, the "risk corridor" could be identified and people warned. The NRC—just like Prudence, Kelvin, and Mrs. N—identifies possible "concerns about property values."

Given a lot of warning and a big budget, space technology might be used to prevent the collision. For bigger asteroids up to 100 meters in diameter, with decades of preparation, *slow push* might be used to put the NEO into a different orbit, which is best done by slowing or speeding up rather than sideways nudges. Using a "gravity tractor"—harnessing the gravitational attraction of an adjacent spacecraft—may be more feasible than actually shoving the rock. Spacecraft have already had close encounters with asteroids: the Hayabusa mission even landed briefly, collected particles, and returned to earth.[11]

With decades of warning, asteroids more than 100 meters and even up to 1 kilometer across might be shifted by *kinetic impact* methods—i.e., ramming with multiple spacecraft. Anything bigger than 1 kilometer across would require hundreds of spacecraft to hit it, or alternatively, a nuclear detonation close by. For 500-meter asteroids, this could be set up in years rather than decades if the political will was behind it.

Anything above 10 kilometers across, the size that wiped out the dinosaurs, is considered essentially unstoppable. Although these apocalyptic scenarios make good, or at least popular, films, the NRC concludes that the main risk is from an unexpected airburst of a small object, less than 50 meters across, but adds: "However, as all

NEOs have not yet been detected and characterized, it is possible (though very unlikely) that an NEO will 'beat the odds' and devastate a city or coastline in the near future." And there's not much you can do about it.

As for man-made junk, about 5,400 metric tons of rubbish has come down over the past 40 years, and there were 28 reentries of satellites in 2013. So far nobody has been injured, even when 40 metric tons rained down on the United States after the Columbia space shuttle broke up. NASA estimated afterward that there was a 1 in 4 chance someone would have been hit. And when the remnants of the Upper Atmosphere Research Satellite (UARS) came to Earth in September 2011, NASA said there was a "one in 3,200 chance of anyone being hit."

But how are such calculations made? The satellite had been up there for 20 years. It stopped working in 2005 and weighed 5,700 kilograms, about the size and weight of a double-decker bus. NASA said it would break into 26 objects that would survive reentry, weighing 532 kilograms in total, about the weight of eight washing machines. These would be spread over about 500 kilometers but cover a total damage area of around 22 square meters (the size of three parking spaces), but they had no idea where the debris would land. As one commentator said, you'd think these geeks would have better control of their satellites—it's not rocket science.

The largest object weighed 158 kilograms, about the weight of an adult gorilla (though that sounds a bit soft—better to think of a couple of washing machines tied together, traveling at 150 kilometers per hour). This does not sound encouraging, but the Earth is a big place, with a surface area of 500 million square kilometers (or 500,000,000,000,000 square meters), and so assuming that the 22 square meters of bits can land anywhere, there is around a 1 in 23,000,000,000,000 (23 trillion) chance that any particular point will be hit.

So if an individual—let's call him Norm—happens to be occupying that point, minding his own business, then, assuming a random landing place, there is around a 1 in 23,000,000,000,000 chance that Norm will be hit—the same chance as flipping a coin 44 times in a row and coming up heads every time, or slightly better than the chance of winning the UK National Lottery twice in a row.

But there are about 7,000,000,000 other people on Earth, and so the chance that anybody at all will be hit is 7,000,000,000/ 23,000,000,000,000 which is 1 in 3,200, which is just what NASA quoted.*

This chance is low, essentially because people don't cover much of the Earth. It may not seem like that when you are up against a stranger's armpit on the subway, but as anyone taking an intercontinental flight will notice, the globe is covered by an awful lot of not-very-much. If each of us claims 1 square meter, that's 7,000 square kilometers in total, which is only 1/70,000 of the Earth's surface. If everyone in the world went to a rock festival in Puerto Rico, which covers about 8,000 square kilometers, they could all squeeze in, although it would be difficult to provide all the accommodations they would need in terms of food service, places to sleep, and restrooms.

The calculation for falling people would be similar. Let's assume one falling dead body, horizontally occupying, say, 2 square meters, every seven years, and let's assume an at-risk area about the size of the London borough of Richmond (about 60 square kilometers, with a population of almost 200,000). This gives us a fairly straightforward, if crude, probability. Norm is comfortable with that. Prudence isn't, it being somehow more real than the end of the world. It works out at about a 1 in 150 chance that one of the 200,000 members of the population will be in the way, every seven years, and, if you happen to live there, a 1 in 30 million chance that it will be you personally, or a 1 in 210 million chance every year.

Is any of this worrying? As ever, it depends on the kind of person you are as much as it depends on the data. In the Lars von Trier's film *Melancholia*, one of the sisters is alarmed by the prospect of doom, the other is relaxed. Trier was reportedly fascinated by a

---

* This analysis assumes that people don't occupy any space. The odds go up a bit if we allow for the width of a body.

comment his therapist made: he said that depressed people often remain calm when confronted by threatening or stressful situations—on the grounds that life is awful anyway. The German philosopher Arthur Schopenhauer made this the basis of a pessimistic view of life in which the only escape from a pointless, eternal failure to gratify human will is through aesthetic contemplation, ideally of music, like Wagner's.

For Norm, who not only isn't miserable enough simply to shrug at existential risk but also feels insufficiently happy for great optimism—he's a middling kinda guy—anxiety about falling objects has more to do with the enormous uncertainty that surrounds the data.

Being average, he should be precisely vulnerable to the average risk of death from an asteroid strike. As he knows, this is the deliciously convenient measure of 1 MicroMort in a lifetime. As he also understands, this is one of the most startling exposures imaginable of the deficiencies of an average, an average that takes in the kind of bolt from the blue—or whatever color space is—so small that it might dent your car or take out a roof-tile, and so rare that the world's media come to take pictures, and that it might possibly finish you off if it happens to choose your 1 square meter among the 500,000,000,000,000 square meters on earth, and he combines this probability with the theoretical probability of total wipe-out all round. In other words, it is almost nothing combined with everything, and then shared out.

To produce from that a lifetime risk for any individual of 1 MicroMort is arithmetically sound but entirely pointless for everyday life. In other words, this average risk tells us almost nothing, even for l'homme moyen Norm. He's not afraid of death, he's afraid for his faith in the power of probability to guide him.

21

UNEMPLOYMENT

"I'm sorry to tell you, Norm, that we're making you redundant."

"What?"

"We're making you . . . "

"Yes, I heard . . . "

" . . . redundant."

" . . . what you said. But I mean no. I mean you can't."

"Norm, you've been a great asset, but . . . "

"No, I mean you can't *make* me redundant. I'm either redundant or I'm not, but you can't *make* me redundant."

"? . . ."

"You can't *make* someone unneeded if they're needed, the condition is predefined, it can't be imposed, it's . . . illogical, completely illogical, it's like saying we're going to make you . . . "

"Norm . . . "

" . . . six feet tall. It's basically blatantly against the law, which means it's not even allowed. And you're meant to consult me anyway, and then if I were to be redundant because the job's not there anymore . . . "

"Norm . . . "

" . . . then you can get rid of me, dismiss me, but you can't *make* me redundant. You can't."

"Norm, we're thinking of giving you the chop because your job's toast and you're scrap. What do you say?"

"Ah. Okay . . . I see."

"Good. You're out."

"Right . . . Yes . . . Got it . . . "

"When?" said Norm.

"How about tomorrow?"

So here he was, coming in to leave. This is what you call a low-probability, high-impact event, he said to himself. In which case, he rather feared he'd miscalculated the probability, as well as the impact.

So. Well. Norm sat at his desk and pulled up his striped socks. He sorted a few papers and put them in the recycling, deleted some emails, made sure one or two people were aware of one or two important bits and bobs and popped in to see what's-her-name in personnel. Norm made a few calls. Norm had a drink at lunch with some of the youngsters, who gave him a nice pen and a card. Norm knocked on the boss's door and said goodbye. "All the best, Norm," he said. Norm put on his duffle coat. Norm walked past the other desks—"Cheers Norm"—and dropped his ID at reception. He walked through the revolving door and stood outside on the pavement.

There had been a couple of occasions in life when he had tried feeling truly miserable, but his heart wasn't in it. Perhaps now? The thought briefly cheered him up. Then that night he dreamed of a man in a duffle coat circling the drain. He hadn't expected it, that was all. And so . . . erm.

. . . .

NORM GOT IT WRONG. He thought the worst would never happen. It's the same error some think wrecked the global finance industry in 2008 and contributed to a deep recession—a failure, as Norm puts it, to take high-impact, low-probability events seriously. How wrong depends on what happens to him next because, at the extreme, unemployment can kill, as we'll see.

Anyone can get a number wrong, especially if you're trying to put a probability on something that never happened before. And because Norm had never been given the elbow until now, he didn't think too hard about it. It wasn't a normal experience.

So the problem for Norm is that normal wasn't reliable. When he calls losing his job high-impact, low-probability, he borrows the language of Nassim Nicholas Taleb in his book *The Black Swan*: "I don't particularly care about the usual," Taleb has said: "If you want to get an idea of a friend's temperament, ethics, and personal elegance, you need to look at him under the tests of severe circumstances, not under the regular rosy glow of daily life. Can you assess the danger a criminal poses by examining only what he does on an ordinary day? Can we understand health without considering wild diseases and epidemics? Indeed the normal is often irrelevant. Almost everything in social life is produced by rare but consequential shocks and jumps."[1]

Some events are easier to put out of mind than others. If they're unusual enough or seem improbable enough, or it's hard to work out how they might come about, is there also a temptation to dismiss them? Although it's not as if unemployment doesn't happen to other people. And it's not as if there haven't been financial crashes before. The point is that we get a little too used to what we're used to. And so this isn't really a failure of calculation (as with asteroids in the last chapter). This time it is a failure of imagination. We can't foresee everything that might go wrong, so we take the easy route of assuming it won't. The remedy isn't just better numbers, it's more varied stories that can take us imaginatively out of our comfort zones—one reason why some people are attracted to the idea of scenario planning as a way of thinking about future dangers.

Still, the numbers could have helped. In 2007 in the United States there were about 7 million unemployed people, for an unemployment rate of about 5 percent, much the same rate as in the United Kingdom. Four years later, after a deep recession, nearly 7 million more Americans and a million more Brits were on the scrap heap, for an unemployment rate of about 8 or 9 percent in both countries.

This is the usual way of saying how bad unemployment is, and by implication the risk that you'll lose your job: simply look up the current rate. Sure enough, this risk went up as the economy went down.

But this way of talking about the risk can be misleading. There are arguments, especially in the United States, about the missing unemployed—those who have given up looking for a job, for example. But that's not the main problem. One statistic helps us to see what's really going on. In the United States in August 2013, according to figures known as JOLTS—because they are from the Job Opportunities and Labor Turnover Survey program administered by the US Department of Labor's Bureau of Labor Statistics—about 4.5 million people found a job. But about 4.4 million lost one, and of these more than a third were layoffs or discharges. That is, they were involuntary, equaling about 1.65 million people who lost jobs they wanted to keep. And this is just one month's data.[2]

What we count when we measure unemployment, and what we worry about when we see the unemployment rate go up, is in large part the cumulative effect of small changes in the difference between the hires rate and the separations rate. But millions in, take away millions out, can equal zero. And when that happens, is zero a good guide to your risk of losing a job? Of course not. So it turns out that the standard way we have for talking about the risk of losing a job typically captures only a tiny fraction of the number of lost jobs. We would argue that this risk is not best measured by the net change in unemployment, or by the stock of unemployed. It is best measured by the total movement of people. This movement can be, and usually is, huge.

The UK data for this movement is still regarded as experimental. JOLTS data in the United States is also relatively new. What it suggests is that when we hear that unemployment has gone up by about 100,000, the number of jobs actually lost involuntarily might typically be of the order of about 16 times bigger. That was the case in both the United Kingdom and the United States. Although, note that these figures include workers who have been hired and separated more than once during the year.

For annual totals the numbers are staggering. According the US Department of Labor, over the 12 months ending in August 2013, hires totaled 52.3 million and separations totaled 50.4 million. Again, about a third or more of the separations were involuntary.

What's happening here is "churn." And it is immense. But these 50 million or more separations yield a net change in unemployment in the United States in 12 months of just 1.9 million. Clearly, the net change in unemployment, like the unemployment total itself, whether up or down, does not capture anything like the full risk of getting laid off. In any large, modern economy, then, the job market is a gigantic revolving door through which millions of people pass in both directions, into work and out, spending varying time on either side.

When we measure unemployment, we typically measure the number of people who are on the wrong side of the door at any one time. But they are only the net change in the stock of unemployed, the millions more who were out of work at that moment, compared with, say, four or five years earlier. These numbers scarcely hint at the immensity of the churn, or the huge numbers who have felt the chill of being on the wrong side at some time or other.

And this is true not only in recessions. It happens in good times, too. US unemployment went down in August 2013, but still about 1.6

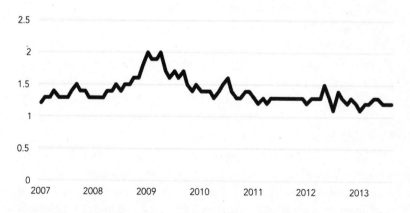

FIGURE 24. Layoff and discharge rates in the United States, 2007–2013: those losing a job involuntarily each month, as a percentage of the total employed.

million people were laid off or discharged. Thinking about the great scale of job churn gives a better measure of the numbers of people who experience unemployment, and is a better way of describing what for many is a state of vast and perpetual risk.[3]

There's another surprise in how this risk changes. Before the recession, the number of layoffs and discharges—the involuntary category—averaged a little below 2 million a month in the United States. This number spiked briefly during the recession, at about 2.5 million, then fell quickly back down. That half-million or so extra jobs lost each month at the peak of the recession in a 140-million-strong civilian workforce in employment converts to a change in the risk of losing your job from about 1.3 or 1.4 percent to a risk of about 1.8 percent in the United States. In the United Kingdom, what the UK Office for National Statistics refers to as the hazard of unemployment changed even less, by about an additional 0.2 percentage points every three months.

An additional risk of 0.5 percent is 1 extra person in every 200 each month. In the United Kingdom, the additional risk of 0.2 percent was equal to about one extra person in 500 every three months. Not nice, but how threatening is that? Yet people did feel threatened. Fear of job loss rises sharply in recessions.

This is a puzzle. Why does unemployment rise so much if the hazard for the average worker doesn't? Part of the answer to the puzzle is that losing your job isn't the whole story about unemployment. Just as significant is what happened at the other end, to the chance of finding a new job if you didn't have one, of getting back through the door once you were out. This is known in the United Kingdom, curiously, as the "hazard of employment."

In the years before the recession in the United States, the monthly hires rate was about 4 percent of the total non-farm workforce. As the recession hit, the hires rate fell to about 3 percent. That 1 percent difference in a workforce of about 140 million people is about 1.4 million fewer hires each month. The hire rate also took a dip in the United Kingdom, where, before the recession, up to a third of the unemployed were back in a job inside of three months.

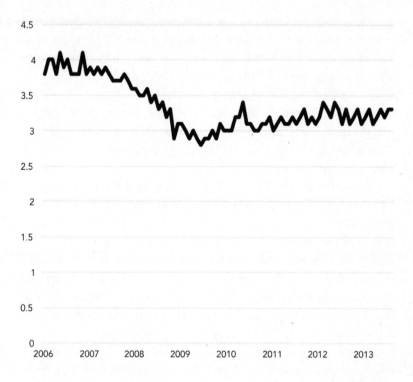

FIGURE 25. The hire rate in the United States, 2006–2013: those getting a job each month, as a percentage of the total employed.

After the recession, this number peaked significantly lower, at about a quarter or less.*

The habit of reporting rising unemployment as if it were all about firms losing business or going bust is misleading; a big part of the story has to do with businesses not opening or expanding. But it is hard to report the nothing of a new job not appearing. It's much easier to show the shutters coming down. Hence "Becker's Law": "It's much harder to find a job than to keep one."[4]

---

* We have simplified the flows by not going into detail about flows in and out of inactivity, the category of people who don't have a job but are not officially looking for work, which we don't think affect the main point of the analysis.

One reason it's harder is that the number of people trying to find work is growing all the time in a rising population. Normally, extra people are soaked up by an increasing number of jobs. This was not often the case between 2007 and 2011. More of the increased work-force stayed on the wrong side of the door.

But it is against the vast, constant labor-market churn that the *extra* threat in hard times of losing your job appears small. To change the metaphor, we are looking at a change in the waves on top of an always heaving ocean.

So the rise in the number of the unemployed is the small difference between huge flows of people in both directions, out of work and into work, a difference owed to the difficulty of finding a job as well as the risk of losing one, plus the extra people wanting work every year; altogether, these factors lower the chance for any individual being employed, sure enough. But amid this great, perpetual flux, the *change* in the risk for the average person of losing a job—if that person already has one—is small. If you thought it was high in the recession, you might well say that it is always high.

Another way of talking about unemployment is to move away from probabilities and say instead that the risk is misery (the probability/consequence distinction rears its head again). According to one study, unemployment is about as bad for our health as smoking. For the young, especially, it leaves a permanent scar on job prospects and wages.

One measure of the consequences is whether unemployment really does become a scrap heap: that is, how long it lasts. In early 2008, about 400,000 people had been unemployed in the United Kingdom for more than 12 months. By early 2012, this number was up to about 800,000. So this risk roughly doubled. Your chance of becoming one of them, even if you did lose your job, was still small, but the consequences could be desperate.

That's something only you can judge. Losing a job might be the chance you were looking for to escape, learn dry stone walling, or cash in the redundancy check and blow the kids' inheritance on foreign travel, in which case the consequential risk is only that you might enjoy yourself.

Or it might be catastrophic. In 2010, the Trades Union Congress (TUC), a federation of trade unions in the United Kingdom, publicized the story of Christelle Pardo as a warning of the costs of unemployment. With no job, on benefits, pregnant, and with a five-month-old baby in her arms, she jumped to her death from the balcony of her sister's flat in Hackney, London. According to the TUC, "her Jobseeker's Allowance had been stopped because of her pregnancy and this meant that she also lost her Housing Benefit: the local authority was demanding that she return £200 in overpaid HB. She had been turned down for other benefits—her appeals had been turned down twice; her last call to the DWP [Department of Work and Pensions] was made just the day before her suicide. Ms Pardo died almost immediately, her son later that day."[5]

Terrible, but a fair reflection of the risk? Losing a job is unlikely in itself to kill you, but the consequences might. Apart from the loss of income, you are also more likely to divorce, commit crime, suffer ill health, and die early. Suicide is often said to be one factor in this rising mortality rate; others are heart disease and alcohol. In short, the unemployed are more likely to be stressed, depressed, broke, and sick.

Estimates of the extra risk of an early death vary widely. One study found no effect; others put it at around 20 percent, with suicide up within a year, and cardiovascular mortality rising after about two years and continuing for more than a decade. A third said the overall increase in mortality as a result of unemployment is nearer 60 percent. That is huge, equal to about an extra 6 MicroLives every day, and similar to the extra risk of being an average smoker.[6]

Is the health effect of a day's unemployment really equivalent to that of a dozen cigarettes, bringing you to death that much sooner, as if every 24-hour day without work left you 27 hours older?

But the problem is not as simple as comparing body counts of the employed and unemployed. Untangling cause and effect in these cases is tricky. Are the unhealthy and unhappy more likely to be unemployed anyway, in which case they might die early for the same reason that they don't have a job, but not *because* they don't have a job? Or is unemployment what makes them unhealthy and unhappy?

One study tried to separate the two by ignoring deaths until a few years after unemployment struck, hoping to weed out anyone who lost a job because they were already unhealthy.[7] It also noted that people who are unemployed because they are sick are usually now on sickness benefits in the United Kingdom and no longer count as unemployed. Its conclusion was that the extra mortality found among the unemployed was overwhelmingly caused directly by unemployment itself.

The size of this fatal effect is still up for grabs, but on current evidence it is probably real. There is also evidence that young people who experience six months or more of unemployment see sharply lower wages in the next few years, and even 20 years later are still earning 8 percent less than those who kept working.[8]

Studies of young people who were unemployed for more than six months in the 1980 and 1990 recessions found that even five years later they spent more than 20 percent of their time unemployed, and 15 percent did so even twelve years later.[9]

High impacts can endure. As Norm has found, risk can take more forms than accidents, and it's not just sausages and cigarettes that can make their effects felt a long way into the future. Nor is it only fatal incidents that can turn out fatally, nor just the known, expected, or calculable dangers that are dangerous. Nassim Nicholas Taleb calls high-impact, unexpected events "black swans," and they are even more troublesome than sometimes supposed. First, the obvious, we cannot predict them; but second, we often don't recognize them when they arrive. You might not know that what you just saw was a black swan until well after it has floated past and either made little difference, proved to be the best thing that ever happened to you, or left your life in pieces, none of which you can be sure of until years later. The 2008 banking crisis was still playing out in 2013, and probably will continue to do so for years to come. Similarly, Norm might not know the risks of unemployment, even after he is shown the door, until years later. So what was the risk at the time that it happened?

22

CRIME

Prudence checked her emails, deleted two financial "opportunities" from Nigeria, turned off the laptop and popped it under the cabinet in the downstairs bathroom, took a last look outside, and then bolted the front door and put on the chain, switched on the alarm, turned off the lights, and went upstairs. She thought again about a dog, especially now that she was alone after her husband had died of prostate cancer, only to run round the same old houses about fleas and mud versus at least a warning bark—and then dismiss the idea. Outside in the wind a tree branch waved into view of the sensor on the security light, switching it on. At the sudden glow in the bathroom window, she twitched, and pulled her robe tighter. It was two years since they had been burgled. She knew already that she wouldn't sleep.

*K2's diary**
No job. Sod it. Go clubbing.
Eight pints = low dosh.
Dosh trip with Kate + fresco shag option.
Old bloke at cashpoint + dog. Dog snarly dog.
Dog snarls. Bloke kicks dog. Snarly dog bites bloke hand.

---

* Kelvin's son.

Big cut. Oozes claret.

Yrs truly also kicks dog.

OB says no one kicks dog, kicks yrs truly knee + language.

Kate bends over helps up yrs truly. Cleavage. Nice.

Old bloke gobs at yrs t, drops dosh.

In consideration of old bloke being knee-kicking/gobbing bastard, and old, shove bloke to ground as he reaches for dosh, grab dosh, tell Kate leg it.

Owing to lack of forethought about efficacy of legging it on bad leg, bloke on floor able to grab bad leg.

With free/good leg kick bloke in head. Language.

Note one foot off ground re kick. Note other foot off ground re grabbed bad leg.

Owing to no feet on ground, fall on bloke.

Dog bites good leg. Bloke bites head.

Pain/claret.

Old bloke fierce bastard. Grinds yrs t face in ground.

Claret.

Kate multiple smashes OB in head with stiletto.

Claret.

OB stamps on hand, grabs dosh, kicks to various body parts inc. stomach, leg, original bad leg, etc., nicks cellphone and wallet inc. cards.

OB legs it + dog. OB nifty for an OB. Conclude later OB probably a crim. Streets not safe.

Kate helps up yrs t again.

Amazing coincidence . . . Emily shows up.

Kate arms round yrs t while yrs t in cleavage close-up scenario. Hi Em.

Em not Hi. Em grabs Kate stiletto, smashes Kate.

Claret.

Em and Kate wrestle on floor. Looks nice. Have idea.

Em bites Kate. Forget idea. Must help. Kick Em.

Kate says no one kicks a woman, which is tech incorrect as just have.

Kate smashes yrs t on head with stiletto. Em smashes yrs t on
head with stiletto.

Claret.

Kate, Em leave together. Spit on yrs t.

Lie in blood/saliva-mix to contemplate strange course of events.
In consideration of deterioration in relationship and recent
facial disfigurement, decide shag option—fresco/other—
prob off.

Stand up v. slow. Go to pub also v. slow. Note unusual coincidence of two limps. Note v. hard to limp with two limps.

Amazing . . . resident pub dodgy geezer offers sale of cash card
and cellphone. Reluctantly decide to buy back own phone.

No dosh.

. . . .

WHICH OF OUR VICTIMS is more typical: Prudence, elderly, alone, and
anxious, or K2, young, drunk, and stupid? Certainly Prudence feels
at more risk in her own home than K2 feels staggering around town
late at night. She's jumpy behind locked doors. He feels indestructible. Reflect on your own impressions of their relative risks and how
they're formed, and we'll come to the answer in a moment.

Meanwhile, here are two ways that you can take the measure of
crime:

1. Examine the numbers
2. Read crime stories in the newspapers

The numbers are imperfect and hard to interpret. Stories are
quick and gripping, full of "psychos," villains, victims and thugs,
inner-city no-go areas, trails of blood and vomit from bar to ER,
robbers on trains, rapists in shadows, fraudsters cloning cards and
conning the old, girl gangs and knife gangs and pushers at the
school gates. All in all, no contest. Asked why they think crime has

been rising for the past ten years in the United Kingdom (when all available data show that it has been flat or tending sharply downward over the long term), people point to stories in the media. Although whether the media cause this perception or respond to it is a moot point. Probably both. In the United States, too, crime has been falling sharply. According to victim surveys, the rate of violent crime since 1993 has fallen about 70 percent, for example, although it has increased slightly in the past couple of years.[1]

We need stories to understand the character of crime. But as with violence against young children, mentioned in Chapter 3, so with other types of crime: dramatic events distort our sense of probability. Read or watch these news stories, and the proof that no one can sleep safely, etc., is roughly one vile crime against a senior or a baby, or a riot. A single instance is about all it takes to confirm that we're all going to hell.

Although it's not the least bit funny if you're a victim.[*] Or if you're afraid—afraid for yourself or for others, afraid to walk down certain streets or afraid in some vaguer way, never quite feeling safe, or afraid at certain times, like alone in bed at night. The President's Commission on Law Enforcement in 1967 was one of the first to say it: "The most damaging of the effects of violent crime is fear and that fear must not be belittled." Surveys consistently report a high-level fear of violent crime (even if they are inconsistent about how they measure it), and no wonder.[2][†]

Fear is one emotional response to perceived risk, not the perceived risk itself, but broadly speaking, the higher the perceived risk,

---

[*] See, for example, the murder of the 94-year-old Emma Winnall, in "Assaulted Pensioner Emma Winnall Dies of Her Injuries," *The Guardian*, May 29, 2012, www .guardian.co.uk/uk/2012/may/29/assaulted-pensioner-emma-winnall-dies.

[†] A common complaint is that the crime statistics lie. Perhaps people have given up reporting crime because they don't think anything will be done. Perhaps the police fiddle the data. But crime figures can also be based not on what people tell the police or on what the police record, but on separate surveys of what people say has happened to them personally. They are approximate, as surveys always are, and they are sometimes updated to take account of new patterns of crime, such as rising theft from children, but they are reasonable.

the higher the state of fear. And we are highly tuned to scare stories in the news, as news editors well know. Fear sells. We're also highly tuned to the latest information and the latest shock story more than to old news or long-run trends. These stories convey fear, not data. And fear is useful, up to a point. It is often good for survival. Unobservant and overly contented never saved anyone from a tiger.

So maybe we are right to grab at every straw in the wind for clues about how frightened we should be, so that we can be careful, again, like Prudence. For the same reasons we are also alert to local rumor—"Isn't the one that lives at number 33 some kind of paedo?"*—or to a few incidents of the same kind of crime close together—"all those knife attacks"—and we take special notice of personal experience—anything really, that stands out, any alarm large and small to trigger our attention and help us to a quick judgment.

The psychologist Daniel Kahneman describes the brain as a machine for jumping to conclusions, working according to a law of least effort. This is especially true about fear.[3] If there's something frightening near you, it pays not to hang around. Kahneman says this mental habit is a cognitive bias, and in our judgment crime stories appeal to it perfectly. Paul Slovic, a colleague of Kahneman's in the 1970s, also showed that vivid events are recalled not merely more vividly but in the belief that there are more of them (see also Chapter 4). And crime is often vivid.

The power of the single example over the mass of data is well studied. As Slovic says, "the identified individual victim, with a face and a name, has no peer." The same goes for animals, he says. During an outbreak of foot and mouth disease in the United Kingdom, millions of cattle were slaughtered to stop the spread. According to Slovic, "the disease waned and animal rights activists demanded an end to further killing. The killings continued until a

---

* A pediatric consultant was mistakenly identified by local, self-appointed vigilantes as a pedophile in the United Kingdom a few years ago, but the mob attack sometimes reported did not occur. See Brendan O'Neill, "Whispering Game," BBC News, February 16, 2006, news.bbc.co.uk/1/hi/magazine/4719364.stm.

newspaper photo of a cute 12-day-old calf named Phoenix being targeted for slaughter led the government to change its policy."

But the plural of "anecdote" is not "data," an old statistical saying goes, and the corollary of "vivid" or "lurid" is not "likely." A crime may be devastating for the victim or family, but it's unlikely to reveal that we're going to the dogs, or to say much about the overall risk of becoming a victim. Isn't this so obvious as to be trivial? Doesn't everyone know it already? Perhaps. But that doesn't stop a massive online tribute by thousands of people to one dead sparrow, shot for knocking over a line of dominoes in a competition, while the Dutch bird protection agency lamented a lack of interest in saving the whole species.[4]

If we would like a slightly more balanced reaction to single vivid stories, perhaps we should aim to make our description of risks so transparent and convincing that it brings "immunity to anecdote." In fact, this notion has been investigated by psychologists studying people's preferences for treatments, who found that good clear graphics, using arrays of icons (as in Chapter 4), could succeed in making people less influenced by stories of wonder cures and ghastly experiences.[5]

Particular events and particular people are usually what stories are about (see Introduction). Not many are about nameless multitudes. In fact, it is often said that without a good dose of the particular, stories lack credibility. Detail makes them feel real—detail that might apply in this instance and no other, a real-world kick in the shins to generality or abstraction. Detail is the handkerchief that Othello gives to Desdemona, and that Iago uses to implicate her in infidelity, small, telling, and "true to life." Detail is the precise story of her murder. It is a claim on believability. The literary critic James Wood has described the importance to fiction of "this-ness"—or "individuating form." By "thisness," he says, "I mean the moment when Emma Bovary fondles the satin slippers she danced in weeks before at the Great Ball at La Vaubyessad 'the soles of which were yellowed with the wax from the dance floor.'"[6] Used well, detail is an assertion that this could happen.

All of which is one way that stories, fictional and anecdotal, differ from the abstraction of probability, which fails the detail test because your own unique circumstances make the average risks that are described by probability hard to apply to you, or to any other individual.

On the other hand, probability does tell us truths of a different order. In particular, it points out that "individuating form" can also fail when extrapolated to other people. That's why it was individuating in the first place. Which is why statistics urges us to learn immunity to anecdote.

These two versions of truth speak a different language, but the problem could also be seen as a simple matter of scale. One version, the truth of the story, gains credibility from being personal and particular; the other, probability, depends on doubting the evidence of a single experience, precisely because it is a single experience, until it is aggregated with everyone else's. In each case, what makes one true makes the other a snake. Are they irreconcilable?

For one of the most extreme crime anecdotes imaginable, take Dr. Harold Shipman. He was what you might call high-risk health care, especially if he knocked on your door for a home visit in the early afternoon. Shipman was a serial killer, a family physician working near Manchester, and as big and vivid a crime story as they come. Reports about suspected victims (named victims, with faces, families, and individuating circumstances) filled the media. A handful of chilling cases gave a stark picture of an avuncular-looking man with a beard inviting the trust of his patients as he murdered them in their own living rooms. It was later estimated that he probably killed upward of 200 people, mostly elderly women in good health, by injecting them with diamorphine.

But for the rest of us he was not a trend, or indicative of the behavior of other physicians, or of anything about the general level of crime. Only in Manchester, where all his murders were recorded as having taken place in one year—the year of discovery—did it make much difference to the ordinary ups and downs in the statistics.

Headlines that play on our deeper fears easily mislead. For instance, murderous stranger-danger to our children terrifies people, but the risk is half that from children's own parents and step-parents. Four out of five adult female victims of murder also know their killer.[7] Alcohol-related violence in England has actually been falling in recent years, contrary to a binge of media coverage, which loves a picture of a guy off his face with a bottle. He exists, it is true, but the news is not a balanced sample of behavior (see also Chapter 4, "Nothing").

But what if lots of stories appear at once, as when four men were murdered—all stabbed—in separate incidents in London on one day, July 10, 2008? This is no longer one shock headline. Here the story was all about how knife killings were becoming an epidemic. We're not talking anecdotes, we're talking data . . . aren't we? Even so, the BBC reported: "Four fatal stabbings in one day could be a statistical freak."[8] Could it?*

Similarly, when NBC News reported in January 2011 that police feared a war on cops, it described what it called "a spate" of shooting attacks: "In just 24 hours, at least 11 officers were shot. The shootings included Sunday attacks at traffic stops in Indiana and Oregon, a Detroit police station shooting that wounded four officers, and a shootout at a Port Orchard, Wash., Wal-Mart that injured two deputies. On Monday morning, two officers were shot dead and a U.S. Marshal was wounded by a gunman in St. Petersburg, Fla."[9]

The annual total of law enforcement officers killed in the line of duty had just gone up 43 percent in a year. You can see why they were nervous. Though in fact, it had gone up from a historic low the year before to something more typical. Even so, the perception that something terrifying was going on was probably real, and no one wants to trivialize homicide. Every incident appalls us. Yet even

---

* Using 2006–2007 as a rough guide, we find that stabbing—or, more correctly, killing by sharp instrument—is the most frequent method of homicide, accounting for 41 percent of incidents in London, with shooting next, at 17 percent. We therefore estimate that four stabbings has a probability of $0.41 \times 4 = 0.028$, assuming that the causes of death are independent for multiple murders on the same day, and that the observed rates can be considered as estimates of the risks for each murder, and given that there are four murders in a day.

the appearance of a spate of killings or shootings might tell us little about the level of crime. Why?

Every murder is an individual crime that can't be foretold. But this very randomness means that the overall pattern of murders is in some ways predictable. This sounds uncanny. It is not (see Chapter 14, on chance). It is probability doing what it does brilliantly, at the large scale. DS asked the UK Home Office how many homicides there had been in London in the past year: 170, they said. So he went away and worked out both how many could be expected up to the current date that year, and the likely pattern of murders over three years, assuming they happened at random. On how many days would there be one murder, two, three, or four, and how often would there be none at all? The pattern he predicted turned out to fit the real data almost exactly.[10]

He denies consulting a crystal ball. But he does admit to using a little basic probability theory.* He did the mathematical equivalent of scattering the murders across the calendar like grains of rice, as if they fell randomly. The result is that four murders in one day (four grains of rice on one space) is unusual but not extraordinary, even assuming that the underlying level of violence remains the same. No new or terrifying trend is necessary to produce this result. We would expect it in London around once every three years. We would expect about 705 days with no murders, 310 days with one, and about 68 days with two. In fact, the numbers were 713, 299, and 66—rather close. We can also work out how often we should expect long gaps between murders—a gap of 7 days should occur around 18 times over the 3 years, and it actually occurred 19 times.

It's no surprise that stories can be unreliable as measures of risk. What this work shows is less intuitively obvious: that so can sudden clusters of incidents. One story is not a trend; nor necessarily are four of the same stories on the same day. Clusters are in this respect normal and predictable and will occur simply by chance.

---

* In fact, the Poisson distribution, which has nothing to do with fish but is named after a Professeur Poisson.

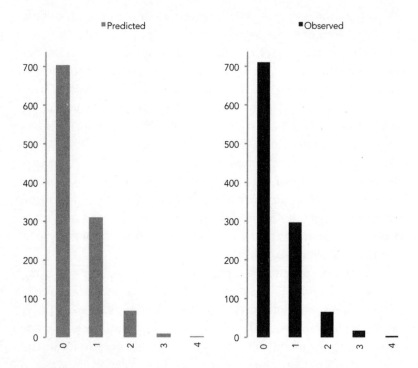

FIGURE 26. Number of days in which there were 0, 1, 2, 3, or 4 homicides in London.

The alternative—a perfectly regular pattern of killings with an equal number of deaths every week—really would be weird. The UK Home Office has since adopted the same analysis to help it check on changes in the homicide rate, commenting that "the occurrence of these apparent 'clusters' is not as surprising as one might anticipate."

On the train home from a meeting at the Home Office at which DS had said we would expect there to have been about 92 murders in London by that point in the year, he picked up a copy of *The London Paper*, which had the headline "London's Murder Count Reaches 90."

So we cannot predict individual murders, but we can predict their quantity and their pattern. And by knowing what pattern to expect, we should also be able to spot when something really un-usual is happening, when the ups or downs are bigger than what

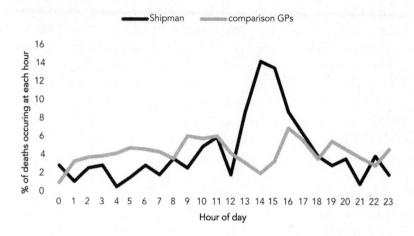

FIGURE 27. Percentage of deaths registered as occurring at different times of the day or night, for Harold Shipman and a group of comparison general practitioners (GPs) in the area.

we'd expect from chance alone. When does the pattern of local burglaries look like chance, and when does it look like a new gang on the block? Four murders in London on one day weren't a surprise, but interestingly, four murders all by the same method—stabbing—probably were. Similarly, the pattern of Harold Shipman's killings told us little about the underlying risk of being murdered in the population as a whole, but with better data at the time we might have been able to spot in that pattern the hugely heightened risk that Shipman was a multiple murderer.

Is Figure 27 the most chilling graph ever? It shows the time of day at which people die, and how those in Shipman's care compared with those cared for by other physicians. Shipman's patients had an improbable habit of dying in the middle of his afternoon visiting hours. Patterns are instructive, if we also know what pattern to expect from chance.

And while detectives examining one death at a time struggled to decide if that particular death was murder, statisticians could use data to discover both the scale of Shipman's murderousness and to tell us where to look for it. And the genius of their investigation was in the simplicity of a question that seeks out patterns among the patients (and the abnormalities in those patterns) by asking, almost trivially: "I wonder what time they died?" The "thisness" of the time of one death in the story of one patient would tell us little in this instance. In aggregate, "time of death" is revelatory, enabling statisticians to deduce far more than the detectives could. Figure 27 has a powerful effect on anyone we show it to. The data reveals a truth that particular stories usually miss, because the truth here is contained in the pattern of repetition in many stories, not in just one of them.

Now that we know to be cautious about shocking but isolated crime stories, and to understand the kind of patterns that occur by chance so that we also know when a pattern is more likely to be meaningful, let's take it a step further and focus on the overall numbers. If these are up, that must mean something.

Which it does, provided the "up" has been long enough and big enough. Otherwise, it is liable to the same problem as clusters. The number of crimes goes up and down anyway, by chance, especially in crime surveys, which are based on a sample that will only approximately reflect what's really going on. No new pattern of criminal behavior, no moral decline or moral revival, is necessary for a certain amount of up and down. "Up" might be chance. "Down" might be a bad sample.

A huge amount of media and political energy goes into short-term changes in the crime figures. If discovering trends is what the pundits and politicians are after, they are often wasting their time. Burglaries up 2 percent this year . . . violent crime down 3 percent, and so on. These statistics seldom tell us anything that couldn't be explained by chance or changing fashions of reporting. A sharp upturn in crime in the United States in 2006, including events reported to the law enforcement agencies and events reported in crime victimization surveys, probably had more to do with a change in statistical methodology than with criminal behavior. Only over the longer term are real changes in crime usually apparent.

Ups and downs also tell us little about the risk of being a victim. "Up 5 percent," but from what? There's abundant psychological evidence that people are more sensitive to change than to base levels of risk. Thirty miles an hour can seem fast or slow depending on what you were doing a few seconds ago. It's the change that we notice more than the absolute level.

But that sensitivity to change can actually distract us from the current risks of being on the receiving end of crime. So finally, what is the underlying, base-level risk of becoming a victim?

Using data, a lot of data, not an anecdote or a cluster or spate of incidents, we can work out the risk of crime simply by taking the number of victims and dividing it by the number of people. And about 30 in 1,000 UK households said they had their homes burgled in 2011, while about 25 per 1,000 people said they suffered violence, though note that "violent crime" in the United Kingdom does not necessarily have to mean injury. Some people experience more than one type of crime, sometimes more than once. Altogether, the chance for the average individual of being a victim of crime in 2010–2011 was about 1 in 5 in the United Kingdom.[11] Fear of crime is not the least bit statistically absurd. Figures from surveys in the United States suggest a similar level of violent crime victimization, at about 25 per 1,000, although violence in the United States tends to be more narrowly defined.

Figure 28 (A and B) gives an indication of the likelihood of becoming a victim of various types of crime in the United States according to crime victimization surveys, which tend to be more representative of the real level of crime than police-recorded crime figures are. Note that the scale for property crime (B) is much bigger than for crimes of violence (A).

But this information is still not that useful. It would have been a fair guide to what might once have happened to Norm, being average, but not for you, not being average, or even for Norm himself now that he is older. By age sixty-five, for example, the risk of violent crime for an individual in the United States is getting down toward a tenth of what it is for a teenager. So the risk for you depends on who you are and which crime you're worried about. For example, those categorized as white Americans are half as likely to

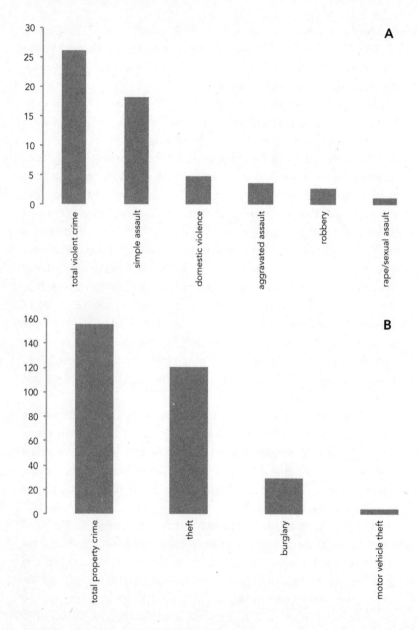

FIGURES 28A AND 28B. Violent crime rates per 1,000 people (A) and property crime rates per 1,000 households (B) in the United States, or the chance of becoming a victim.

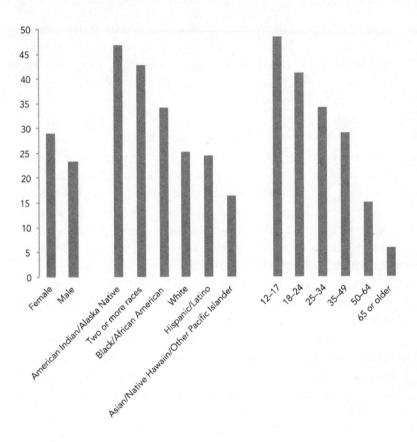

FIGURE 29. Violent crime in the United States by victim's sex, race, and age, rates per 1,000 population.

be victims of violent crime as those described as American Indian or Alaska Native. So who is at risk of what in the United States? Figure 29 shows who tends to have the greatest chance of becoming a victim of violent crime in the United States. Based on these probabilities, the Prudences of this world ostensibly have a lot less to worry about than the Kelvins.

It also matters where you are. Strangely, you were about 16 times more likely to be murdered in 2010 in Cumbria, a county in England best known as the Lake District, than in Dyfed Powys in Wales, but that turns out to be the cluster problem again. The grim fact underlying the point was that in Cumbria that year the number of homicides went up dramatically when one man, Derrick Bird, shot and killed 12 people. There is always a risk of one-offs, but one-offs do not determine the risk of crime.

A more reliable geographical indicator is that almost half the reported robberies in England are in London.[12] Another arises with the fact that some types of crime are far more common in poorer areas. Again, what makes some patterns instructive and others not is when they rise above the levels of variation that we might see from chance alone. Altogether, a youngish bloke like K2 out and about at night is more representative of crime victims than getting-on-a-bit, middle-class, married Prudence at home in the suburbs. K2's night-club habit is surprisingly relevant, although is it his age that is the real risk, or the fact that young people are more likely to go to night-clubs, that helps explain why youth seems to be relatively dangerous? Either way, all in all, Kelvin is a walking risk factor.

For obvious reasons, homicide doesn't appear in crime victimization surveys, so we rely on incidents known to the law enforcement agencies. The underlying rate of murder and non-negligent manslaughter in the United States, according to the FBI, is 4.7 per 100,000.[13] Comparing this with the UK rates, or indeed, comparing almost any set of crime rates, is treacherous, often impossible to do accurately. Some say Britain is far more violent, some say America is.

Serious definitional differences abound, not only between countries, but also within them. Violent crime in the United States, for example, tends to mean physically violent crime, as we have said. But in the United Kingdom, it means just about any offense against the person, and can include harassment or causing distress. Which is one reason UK police-recorded violent crime rates appear higher than American police-recorded violent crime rates (rates in surveys are more similar). But when about half of UK violent crime does not involve any physical harm, comparisons of headline recorded crime rates are clearly pointless. Similarly, at first glance it looks as if the US murder and non-negligent manslaughter rate of 4.7 per 100,000

(47 MicroMorts a year) is nearly four times the UK rate of about 1.2 per 100,000 (12 MMs per year). And all Brits know, or think they know, how violent America is. But a chunk of this difference could be due to reporting practices: in the United Kingdom murder is only murder when the courts say so; in the United States, if the police suspect murder it is counted as murder. In our judgment, murder probably leaves less room for definitional doubt than some other categories of crime, not least because a dead body that came to a sticky end sure narrows the possibilities. A difference of four times greater is a lot to lose in the definitional margins for this type of crime. But what the real difference is, we wouldn't like to say.

Whatever you think about the international comparison, these average risks of murder and non-negligent manslaughter in the United States of 47 MMs per year are still tiny when compared with most average risks in life. We adopted the MicroMort partly for its small scale, so that it can pick up small risks, but on this individual scale the average risk of becoming a victim of homicide is about as close to negligible as a real risk can be.

But who's average? Prudence for one would not be reassured.

23

SURGERY

The patient was in a bad way: white, male, about 185 pounds, history of heart disease, complaining of chest pain and shortness of breath.

"Blood pressure 85 over 60, falling," said a nurse. "Breathing erratic."

Pulse? Where was the damn pulse?

There was no time. No time for tests, no time to lose. The brilliant but unorthodox surgeon, Kieran Kevlin, fifty years old, silver-haired twin sibling of a famous former professor at the Sorbonne and at the peak of his powers, knew he must operate—at once. Hemorrhage? A valve? His mind raced.

"Can we save him, doctor?" breathed Lara, the surgical nurse, as her gloved and skillful hands prepared the patient. A wisp of blonde hair fell across those anxious but still beautiful features beneath the green surgical mask.

"It'll be touch and go," said Kieran, looking deep into her compassionate blue eyes, his square jaw set with the steely resolve she had long secretly loved. "But for his sake, for his family's sake, for the pride of this hospital and the values we share, let's give it our best shot."

From around the operating theater came murmurs of assent. Kieran was known as a maverick, but also a medical virtuoso, a man who could play ducks and drakes with a battery of surgical techniques before breakfast and follow it with a ten-mile run.

"Thank you, doctor," said Lara, laying her hand on his gown and sensing the muscular forearm beneath. "You might never know how much we all admire your . . . your . . . "

"No, no, thank *you*, Lara. I appreciate your always beautiful thoughts. But there's no time for that now. We have a life to save."

And although he knew the gravity of the crisis, this was also a moment he lived for: the moment of decision and incision, the test of self-belief. To put a knife into human skin, deeply aware of his own fallibility, naturally, and of all that could go wrong by ill luck or bad judgment, but counterbalanced by training to perfection, even if he did say so himself, which luckily combined in his case with natural genius. He was born to be a savior of broken bodies, an artist reveling in his craft, a musician transcendent in his calling, primed to wound that tender flesh with a razor-sharp instrument, to make that first cut and see those delicious, delicate red pearls, to re-arrange, carve up, butcher, slash, and rip and burn and slice like ripe mango, oh yes, and stitch and restore and make whole and *re-create* a God-given human body—this was life and meaning.

He made the cut, swift and deep, and then glanced at Lara to see her soft and yielding blue eyes still upon him. He felt her faith. He could not let her down. Yet he also knew that it would be a damn near thing. He was on the edge, trusting to instinct. Even so, he smiled that roguish, dimpled smile of his, and winked.

Lara's heart filled with desperate joy. If only, she thought, if only he were not so good and devoted to his wife and family. But no sooner had the thought come to her than she felt ashamed—and cursed herself for so selfish, so hurtful a dream. It was not to be. It could not be. She would never know happiness, except for the happiness of being with him now, seeing him in his element, saving lives.

When finally Kieran left the patient to a colleague to clean up, trusting to a smooth recovery, and when his hands met Lara's as they took off their gowns, and they simply stopped, stood, and stared into one another's souls for what seemed like eternity, and shortly afterward somehow fell into the back of his car in a far corner of the parking lot, where their first lovemaking was of a sweet intensity that strangely reminded her of his surgery, under the skill of his strong hands, so that the very notion of guilt seemed unkind for an act of such brief perfection—extremely brief, actually—and then when, the very next day, he was suspended, and (not long after she realized she was pregnant) later sacked, and finally struck off the medical register for negligence, following the patient's death for a procedure subsequently condemned as "ludicrously ill-judged," and, to quote the serious-incident report in the ensuing litigation, "cavalier in its willful disregard of basic medicine and good sense, as if Mr. Kieran Kevlin had been more motivated by the daredevil swipe of the scalpel," it seemed to her as if a brave and glorious bubble had popped.

. . . .

THE BRITISH DOCTOR-PHILOSOPHER Raymond Tallis says he is privileged beyond the dreams of his ancestors, thanks to modern medicine. That's a bold but fair claim, and Tallis is a superb advocate for the benefits of health care.*

Medical drama goes even further. Doctors in fiction are saviors. Sick people are pulled through by techno-miracles or genius. If patients die, the failure is heroic or inevitable. On the whole, our fictional doctors don't kill us slicing out the wrong bits.

Although when a real doctor ordered that an epilepsy drug— phenytoin—be given to Bailey Ratcliffe, a five-year-old boy with a bad fit who seemed unresponsive to other drugs, she got the dose wrong by a factor of about 6. He died. "I'm sorry," she said at the inquest in December 2012, "I made a mistake."

Doctors do. For all medicine's heroics, and for all the genuine, extraordinary good that it does, white-coat confidence like Kieran's

---

* See, for example, his lecture "Longer, Healthier, Happier? Human Needs, Human Values and Science," 2007, Sense About Science, www.senseaboutscience.org /pages/annual-lecture-2007.html.

has not always been as well earned as doctors like to believe, or their patients like to think. So here's a question: Does popular fiction about life-and-death risks in medical drama reflect the data? Or, by taking on a life and conventions of its own, does it distort popular and professional belief about the dangers?

Kieran is a calculated insult to medicine, partly—wickedly—just to see how it sounds, a parody of the medical hero who kills people by accident and incompetence. It's not the way the story is normally told, and we're not suggesting Kieran is the norm, but should the medical story be told that way more often?

Medicine today is coming around to a wider acceptance of error and uncertainty and is becoming more willing to acknowledge risks from mistakes and ignorance about what really makes us better, or not. As a result, it is making more of the progress Raymond Tallis describes. But it can still be guilty of what the British Medical Journal once satirized as seven alternatives to evidence-based medicine, among them: "eminence-based medicine," where the more senior the colleague, the less need for anything so crude as evidence that the treatment works; "vehemence-based medicine," or the substitution of volume for evidence; and "eloquence-based medicine," for which "the year-round suntan, carnation in the button hole, silk tie, Armani suit, and tongue should all be equally smooth."[1]

The growing popularity of statistical hard graft to find out who makes the most mistakes, or if people really get better, or worse, because of how we treat them, and how much better, is surprisingly recent. The Journal of the American Medical Association announced the arrival of "Evidence-Based Medicine—A New Approach to Teaching the Practice of Medicine," which "de-emphasizes intuition [and], unsystematic clinical experience," only in 1992. One critic still maintains that most published research findings are false, because studies get it wrong, and yet the ones that seem most exciting tend to be those that are published. All of which hardly puts your mind at ease if you're about to go under the knife.[2]

Some medical dramas have picked up this humility. One in particular—the American comedy Scrubs, which follows a group of all-too-imperfect new doctors*—was inspired in part by a compelling account of medical error by a US surgeon, Atul Gawande, in his book *Complications*.[3]

Gawande is fascinated by fallibility. His books burst with medical mistakes, and he readily admits his own, including botching the insertion of a central line into the main blood vessel to a patient's heart. In Gawande's version of the medical narrative, screwups, large or small, are routine, even necessary to medical training.

"The stakes are high, the liberties taken tremendous," he writes. "What you find when you get in close, however—close enough to see the furrowed brows, the doubts and missteps, the failures as well as the successes—is how messy, uncertain, and also surprising medicine turns out to be."

He describes it as an imperfect science, "an enterprise of constantly changing knowledge, uncertain information, fallible individuals and at the same time lives on the line."

But is that the public perception? Or is our mental model dominated by simple stories and ideas of treatment and cure? If the latter, we might underestimate the risks. Hence our story. We wanted to play with the narrative tradition by giving the hero feet of clay, a sack of moral failings, and a huge error of judgment.

So, having felt Kieran's disgrace, do you feel any different about medical risks? Probably not. His is just one story. And to be credible, stories need—as lawyers say in cases of defamation—a substratum of provable fact. That is, they also need evidence, or at least belief about what the facts and data really say.

As ever, then, some facts and data to put into the mix of perceptions and stories.

Surgery is simple. A human body is soft, so it takes only a sharp knife to slice out giblets and a saw to hack bits off. The complication is how to stop the patient from dying through blood loss, agony, infection, etc. Given what we now know about these hazards, it's hard

---

*In one episode of *Scrubs*, the "hero" is shown on his rounds with a ghost in tow—the ghost of a patient whose death he has caused by medical error.

to read about surgery in the past without flinching: at the crudity of the tools, the lack of hygiene and anesthetic, and, not least, the insane ambition.

Take trepanation, in which part of the skull is removed to reveal the brain, once widely practiced either as relief for headache or following injury. The head was particularly prone to damage from slings and clubs and other primitive weapons. The aim of trepanation was to relieve what felt like extreme pressure, release blood and "evil air," and leave the brain nicely aerated.[4]

Excavations reveal that in Neolithic times as many as one skull in every three had holes drilled or scraped out. The even more remarkable finding is that the original owners of many of these skulls—between 50 and 90 percent, according to some sources—survived. We know this because the edge of the hole has healed. The procedure was popular in Europe as a treatment for epilepsy and mental illness up to the eighteenth century, and afterward for head injury. Miners from the English county of Cornwall in the nineteenth century apparently insisted on having their skulls bored as a precautionary measure after even minor head injuries.

But it was when hospitals took over that holes in the head became especially dangerous. The problem—as with maternal mortality, as described in Chapter 11—was hygiene, with the infection risk inside a hospital so high that doctors managed to take a mad idea and make it worse. The mortality rate shot up to about 90 percent. Once again, professionals and institutions were bigger killers even than the procedure itself, taking out perhaps an extra 80 percent of their patients. Which is why the high numbers of survivors found in nineteenth-century excavations seemed so unbelievable. How could ancient Peruvian natives carry out this operation successfully? Like giving birth in the nineteenth century, it was far safer to have a hole drilled in your head at home.

Other than having your head excavated with a sharp stone, the only pain relief available for so-called primitives was intoxication. Alcohol, cannabis, and opium were the basic anesthetics until Humphry Davy personally experimented with nitrous oxide, or laughing gas. In 1800 he had the foresight to write: "As nitrous oxide in its extensive operation appears capable of destroying physical pain, it may probably be used with advantage during surgical operations in which no great effusion of blood takes place."[5] Naturally, nobody in medicine took any notice for fifty years, while laughing gas and ether were used as party tricks. "Ether frolics" were hugely popular in the United States. At last it dawned on some medical students that the frolickers appeared not to care about injury. Could this be put to practical use? they wondered.

The first public anesthetic using ether was delivered by William Morton on October 16, 1846, at the Massachusetts General Hospital. The idea soon spread, especially after Queen Victoria grasped at chloroform for the birth of Prince Leopold in 1853, although chloroform later lost favor owing to sudden deaths from heart arrhythmias, now known as "sudden sniffer's death" among teenage solvent abusers.

To be numbed and put to sleep for an operation is now routine—the World Health Organization reports that each year some 230 million major surgical procedures are conducted with patients under anesthesia—with rates strongly dependent on health-care spending. Anesthetics are fairly safe now—as noted earlier, the American Society of Anesthesiologists says that, "at present, the chances of a healthy patient suffering an intraoperative death attributable to anesthesia is less than 1 in 200,000 when an anesthesiologist is involved in patient care." That's a risk of 5 MicroMorts, equivalent to around 20 miles on a US motorcycle, or 1,000 miles in a car, or around that of a scuba dive. Risks for outpatient surgeries are lower, but higher if you are older or it's an emergency operation.[6]

Anesthesiologists are fond of saying that the risk is less than that from driving to the hospital for the surgery, which is only generally true if you come on a motorcycle or are a spectacularly reckless driver. If the rates claimed in the United States applied to all the 230 million operations reported by the WHO each year, this would mean 1,150 deaths by anesthesia—almost certainly a big underestimate.

And hospitals can hurt you in other ways than through surgery, whether it's through getting an infection or through slipping on a pool of something unpleasant. In 2010, there were around 35 million hospitalizations in the United States and 715,000 deaths: that's around 20,000 MMs per admission.[7] Of course, many of these deaths were inevitable, and the proportion that were "avoidable" is unclear: estimates of around 200,000 are common, but these include adverse reactions to correctly prescribed drugs and infections acquired in hospitals as well as classic "medical errors" and accidents.

So Kieran's complications story conveys a truth about hospitals: they remain dangerous to our health even while helping it beyond the dreams of our ancestors.

Risk in medicine is unavoidable. But the scope for error adds to it, as does plain bad luck. So it is no surprise that the risk of an operation varies between hospitals and surgeons. The concept of measuring their performance began with Florence Nightingale: after tackling the squalor of military hospitals during the Crimean War against the Russians in 1854, she was keen to do the same in England. Obsessed with statistics and a passionate admirer of Quetelet, she viewed the patterns in the data as an indication of God's work. To study them was a spiritual endeavor.

Nightingale proposed the collection of "uniform hospital statistics" to "enable us to ascertain the relative mortality of different hospitals." But she was aware that hospitals were adept at fudging the figures by dumping hopeless cases onto someone else: "We have known incurable cases discharged from one hospital, to which the deaths ought to have been accounted, and received into another hospital, to die there in a day or two after admission, thereby lowering the mortality rate of the first at the expense of the second." Nowadays we call this practice "gaming." The Victorians could be ruthless in concealing poor performance, just as some hospitals are now, as various scandals suggest, and her grand plan fizzled out.[8]

Forty years after Florence Nightingale, the Boston surgeon Ernest Codman took a different approach to checking the quality of care. Rather than publishing overall statistics, his "End Results Idea" required hospitals to complete a small card for each patient that explained publicly and in detail whether the treatment was successful and if errors were made. He began this himself from 1900, and even opened his own private hospital in 1911. He claimed that his ideas "will not be eccentric a few years hence,"[9] and, unlike Nightingale, he courted controversy, causing an uproar at a public meeting by unveiling a huge cartoon satirizing the Boston medical establishment for carrying out expensive and unproven procedures, and so grabbing the "golden eggs" laid by an ostrich representing a gullible public. His scheme, unsurprisingly, did not catch on. Harvard fired him as surgery instructor, and his hospital closed in 1918.

There have been modern attempts to emulate Nightingale and Codman, notably for heart surgery, but the quality of the data about hospital performance is probably still not as good as the public thinks it is. Let's take a closer look at one of the exceptions, where the data are not bad, and with a little nosing around we can begin to discover the limits of what we can know about medical risk. This is the coronary artery bypass graft, known as a CABG (pronounced like the green vegetable). CABGs are intended to relieve angina by improving the blood flow to the heart with a piece of artery or vein taken from elsewhere in the body. This type of operation started in the 1960s, and mortality in the United States was down to 3.9 percent in 1990 and to 3 percent in 1999. The United Kingdom now reports a "98.4 per cent survival rate," based on 21,248 operations in 2008.[10]

Note the different framing of the information in the United States compared with the United Kingdom. In the United States, people die from surgery, while in the United Kingdom they do not survive. This change of framing is a neat device that tends to make performance look better and obscure differences: the difference between two hospitals with 98 and 96 percent survival, as we would describe them in the United Kingdom, looks negligible, while the same comparison expressed as it would be in the United States, as 2 percent versus 4 percent mortality, looks like double the trouble.

Some states in the United States mandate mortality reporting. For example, all hospitals in New York State that perform cardiac surgery must file details of their cases with the State Department of Health. In 2008 there were 10,707 CABG operations in 40 hospitals, and 194 patients died, either in the hospital or within 30 days—making for a mortality rate of 1.8 percent. Or, as they would say in the United Kingdom, a survival rate of 98.2 per cent.[11]

Surgery on heart valves was a higher risk: of 21,445 operations between 2006 and 2008, 1,120 patients died—a mortality rate of 5.2 percent, or just over 1 in 20. That's an average of 52,000 Micro-Morts per operation, equivalent to around 5,000 parachute jumps, or two RAF bombing missions in World War II. Needless to say, this is serious, but presumably the judgment is that the risk without an operation is higher.

This is an example of where we have data: we can define the risks, and that allows us to make comparisons between hospitals. But how useful are they? That sounds like a daft question. If they are real data, what could possibly be wrong with them?

What follows is a trip along the slippery path of working out what the dangers really are for a group of hospitals—even based on relatively good data. Stick with it if you can. It is an object lesson in the difficulty of coming to clear conclusions about risk, and another reminder of medicine's uncertainty about how good it is really.

Taking the numbers at face value, you might prefer the hospital with the lowest rate of deaths, which happens to be Vassar Brothers Medical Center in Poughkeepsie, New York, which had only 8 deaths in 470 operations (1.7 percent). At the other extreme, we find University Hospital in Stony Brook, which reported 43 deaths in 512 operations (8.4 percent). But would you be right in this choice?

Maybe Stony Brook was treating more severe patients. It was for this reason that Florence Nightingale decided 150 years ago that crude mortality rates were unreliable for comparisons, since hospitals differ in their "case-mix." Ever since there have been attempts to "risk-adjust" the data to check whether the difference in the number of deaths could be accounted for by the type of patient.

In New York they collect data on the age and severity of the patient's illness and build a statistical equation that tries to say what the chances are that each patient will die in an "average hospital." For the type of patients treated in Stony Brook, we would have expected 35 deaths, compared with 27 if they were treating only average patients, which shows that Stony Brook really did have a tendency to treat older or more severely ill patients.

But, as we mentioned above, Stony Brook actually had 43 deaths, 8 more than expected, so even the case-mix does not seem entirely to explain the high number of patient deaths. With 43 deaths rather than the expected 35, we can say the hospital had 43/35 = 123 percent of the expected mortality. New York State Department of Health takes this excess risk of 123 percent and applies it to the overall death rate in New York, 5.2 percent, to give an overall "Risk Adjusted Mortality Rate" of 123 percent of 5.2 percent, which is 6.4 percent, a figure intended to reflect the risk for an average patient treated in that hospital (see Figure 30).

But even a good surgeon might have a run of unexpected bad cases. Kieran's patient dies, but was he just unlucky? Did he have one uncharacteristically off-day, distracted by the lovely Lara? Was Stony Brook just unlucky to have a mortality rate (risk-adjusted) of 6.9 percent, compared with the statewide average of 5.2 percent? The question is becoming increasingly sophisticated: How do we measure bad luck?

Fortunately, statistical methods are good enough to have a stab at it. They depend on making a distinction, mentioned back in Chapter 1 ("The Beginning"), between the observed mortality *rate*—the historical proportion who died—and the underlying mortality *risk*, which is the chance a similar future patient has of dying. Rates will not exactly match risks, just as 100 coin flips will rarely end in exactly 50 heads and 50 tails. There is always the element of chance, or luck, or whatever you want to call it.

We can check the role of the ordinary ups and downs of good and bad luck by using a *funnel plot*, which shows the mortality rates for all 40 hospitals plotted against the number of patients treated. Smaller hospitals are to the left and bigger ones to the right. The areas inside the "funnels" show where we would expect the hospitals

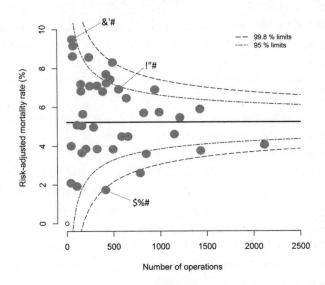

FIGURE 30. "Funnel plot" comparing the "risk-adjusted mortality rates" for New York hospitals carrying out adult heart surgery. If a hospital lies outside the funnels, there is reason to think its mortality rate is truly different from average. Vasser, Stony Brook, and New York Hospital Queens are labeled.

to be if the actual risk for each patient was equal to the overall average and the only differences between hospitals were due to chance. The funnels are wider for the smaller hospitals because they treat fewer cases, and so a bit of bad luck can make a bigger difference to their apparent performance. If it were really the case that they were all average and the only difference between them was luck, 95 percent of hospitals (38 out of 40) should lie in the inner funnel and 99.8 percent (40 out of 40) should lie in the outer funnel.

In fact, five hospitals lie above and five below the 95 percent funnel, eight more than we would expect by chance alone; and two even lie below the 99.8 percent funnel, suggesting they are doing surprisingly well. Stony Brook (SB) lies well inside the funnel, suggesting that its apparent excess mortality could be entirely due to bad luck, and that there is no reliable evidence that it is an unusually dangerous place to go for treatment. Vassar (VA), on the other hand, has an extraordinarily low number of deaths, even allowing for its case-mix. It looks genuinely good.

The "worst" hospital is apparently New York Hospital Queens (NQ), with a risk-adjusted mortality rate of 9.5 percent, almost double the statewide average. But this is based on only 6 deaths in 93 operations, and so we cannot be confident that this was not just a run of bad luck—it is also inside the funnel.

The United Kingdom now produces funnel plots of a Summary Hospital-level Mortality Indicator (SHMI), which compares hospitals according to how many patients die within thirty days of admission, adjusting for the type and severity of their condition.[12] This measure is controversial. For example, what about very sick patients admitted for palliative care only because they are expected to die? Unless allowance is made for these, perhaps hospitals would start using the Nightingale trick of either refusing them admission or rapidly getting rid of patients that were too sick.

One approach has been to allow hospitals to tick a box saying the patients were admitted for palliative care, but the temptation has been to do this for as many as possible, as it raises the "expected" numbers of deaths and so makes performance look better. At the extreme, there is evidence that some hospitals coded as many as 30 percent of their admissions for palliative care.[13] This box has been removed from the current system.

Measuring hospital risk is a fascinating example of the genius and limitations of statistics. The risks at each institution ought to be calculable—and to some extent they are. But although we can measure differences, and although we can even estimate the extent of luck and bad luck, and though we can use techniques such as the funnel plot to picture all these data and put them in context and show them so that they are easier to understand, we cannot be sure, just as Florence Nightingale could not be sure, that we have eliminated human ingenuity for playing the system. The people factor is still there to mess up the odds. If you want to know the safest place for your operation, check the statistics, but don't expect a simple answer.

24

SCREENING

Should she, shouldn't she? Prudence was seventy. The invitation—to be screened for breast cancer—lay on the table. It came with a leaflet that told her, more or less, to go; it might save her life. And in the past she had. But now there were these rumors. God, did they know how to make an old woman suffer.

"It's about three to one," said Norm, offering advice.

"It's well after two," said Prudence.

"The ratio," said Norm.

"Oh yes," said Prudence, "the ratio."

"Of the number of people who are treated unnecessarily when the lump is harmless, compared to the number of lives saved."

"And if it's a positive test, does that mean I'm the one or the three?"

"Can't tell. If we knew that, it would be easy."

"And when would we know?"

"Never."

"Never? Even after they'd chopped them off?"

"Even after . . . [staring somewhere else] . . . so it's for you to decide, really, how to weigh the balance between a life saved and three times as much chance of an overdiagnosis with . . . erm . . . collateral damage."

"Yes, Norm. But I'm frightened. Even if I'm old."

"Oh," said Norm, looking down. "Maybe you need to learn to live with uncertainty?" he said, looking up, "and relax."

"Relax?"

"Relax." He smiled.

"I don't want to play the odds, Norm. It's too late. I want not to be afraid. Please, how can I not be afraid?"

"Ah, erm . . . erm . . . "

. . . .

FEELING ILL? "Fine, thanks," you say. Not worried that the twinge in your chest might be . . . something. "Never felt better," you say. Ah, then perhaps this well-being hides a risk yet to strike, an unseen killer, lurking in your genes or blood. Are you sure you wouldn't like some professional reassurance?

"Well, now you mention it . . . "

Reassurance—peace of mind—is often the health industry message. It's what Prudence wants. And screening sounds like a good way to get it. The impulse to "find out," to "check," which Prudence feels powerfully, imagines a day when doubt is put to rest. And it's easy nowadays to find clinics to examine and scan us for a worrying range of diseases that we might have without realizing it. There are effusive testimonials from smiling people who have been "saved" by these tests. What could be the harm in having a checkup? Possibly, quite a lot.

The story of screening taps into the story of a medical cure that we looked at in the previous chapter, and goes like this: Woman (Prudence) is worried. Woman goes for checkup and screening. Screening discovers cancer. Cancer is treated. Woman is saved. Screening saves women's lives.

There's a simple linearity here of cause and effect. But is it the only one? Is it, that is, the right story? Prudence is suddenly not so sure, and having her reassurance taken away is hard. She feels this because of recent reports that there is another story about screening—that it also, sometimes, causes harm. The big problem is that we don't know who will be harmed and who will be saved. So in some ways screening creates new kinds of uncertainty and new threats of harm.

How? And how much harm compared with the life-saving benefits?

Consider another screening system: security. Here's something to think about as you inch forward in the queue to be allowed into someone else's country. Suppose you've rounded up 1,000 of the usual terrorist suspects, they all declare their innocence, and someone claims to have a lie detector that is 90 percent accurate. You wire them up, and eventually the machine declares that 108 are probably lying. These are taken away, given orange suits, and not seen again for years. No doubt one or two could be innocent, but it serves them right for being in the wrong place at the wrong time.

But then as the years pass and the court cases for false imprisonment mount, you begin to wonder about this 90 percent accuracy. You go back and check the small print, which says that whether someone is telling the truth or a lie, the machine will correctly classify them in 90 percent of cases.

But this means, believe it or not, that most likely there were just 10 terrorists in the 1,000, even with "90 percent accuracy." The sums are fairly basic: the test would pick up 9 out of 10 of the real terrorists, letting 1 go free. But there were 990 innocent people, and the test would incorrectly classify 10 percent of them—that's 99—as "terrorists." That makes 9 + 99 = 108 people sent off to a remote prison, 99 of them wholly innocent—that's 91 percent of the accused wrongly incarcerated by a "90 percent accurate" test.

You may think this story is exaggerated, but this is exactly what happens in screening for breast cancer using mammography: only 9 percent of positive mammograms are truly cancer, and 91 percent of the apparently "positive" results—many of which will cause a good deal of anxiety to women, some of whom will go on to have biopsies and other investigations—are false positives. The 10 percent inaccuracy bedevils an awful lot of healthy people.[1]

Mammography is quite good as screening tests go, as it correctly classifies 90 percent of cases, but because only around 10 in every 1,000 women screened have breast cancer, most of the positive results are in people who don't have it: false alarms. It's like looking for a needle in a haystack, when a lot of the hay looks like needles, and explains why the vast majority of people who set off airport security alerts are innocent.

Of course, maybe delaying some passengers, giving some anxiety to women, and locking up some suspicious-looking people may be a price worth paying. But this is not the only problem with screening tests.[2]

Next, there is the possible harm of the test itself. If we believe the linear no-threshold principle for radiation damage (see Chapter 19), then we can estimate the harm done by imaging healthy people. We've seen that CT scans are estimated to cause thousands of cancers, but most of these will presumably be for some diagnostic purpose. Airport scanning is more controversial, and each "backscatter" X-ray is limited to 0.1 micro-Sieverts (1 banana). This is only equivalent to a few minutes of background radiation, and is around 1 percent of the exposure from a five-hour flight itself. A frequent flyer who had 4,000 of these scans would be exposed to radiation equivalent to 1 mammogram, and if we really believe the LNT principle, then 100 million frequent flyers would end up with 6 cancers caused by the scanning, as part of the 40 million cancers they would get anyway.[3]

The radiation harms of mammography have been estimated as follows: the British NHS Cancer Screening Programme reports that if 14,000 women are screened for 10 years (in Britain women are offered screening every 3 years, so this is 3 mammograms each), around 1 fatal cancer will be caused. If we assume that each fatal cancer costs a woman 20 years of life, then that is around 8 MicroLives per mammography, around 16 cigarettes, exactly the figure we arrived at in Chapter 19, when discussing radiation.[4]

A recent US study puts the risk somewhat lower, estimating that if 100,000 women are screened every other year between the ages of 50 and 59, this will lead to 14 cancers and 2 deaths, although the recommended US screening regime would lead to 86 cancers and 11 deaths.[5]

But the main problem with medical screening is what is called overdiagnosis or overtreatment, or harm done by medical care itself.

It's a simple idea: things are treated that would never have been a problem anyway.[6]

This is the problem that preoccupies Prudence, and it was recently quantified by an independent report in the United Kingdom, which stated that for every woman whose early death from breast cancer was prevented by screening, three women were given full treatment for a breast cancer that would never have bothered them if they hadn't gone to the screening. This information is now included in the official leaflet sent to women with the letter inviting them to breast screening. The leaflet is a remarkable innovation, in that it does not actually recommend attendance at screening, but only lays out the possible benefits and harms in order to encourage an informed choice.[7]

The curative screening story we told at the beginning was entirely about events, about the bad things that happen and how medicine can save us (see Chapter 4, on non-events). But it is a story that

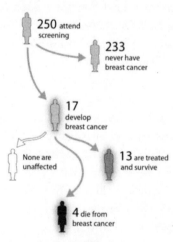

250 women between 50 and 70 who are not screened

- **250** are not screened
- **233** never have breast cancer
- **17** develop breast cancer
- **3** are unaffected
- **9** are treated and survive
- **5** die from breast cancer

3 fewer treaments, 1 extra death

250 women between 50 and 70 who attend screening

- **250** attend screening
- **233** never have breast cancer
- **17** develop breast cancer
- None are unaffected
- **13** are treated and survive
- **4** die from breast cancer

3 more treaments, 1 fewer death

FIGURE 31. The risks of screening and not screening for breast cancer between ages 50 and 70 with follow up until age 80.

can go wrong, both in detection of the bad events and by sounding false alarms over non-events, and also even by causing bad events. People who pass through the system follow many different paths and have many different stories, as Figure 31 shows.

So although Prudence wants reassurance, and screening looks like the way to get it, it is in some ways back to the classic problem with risk: it tries to define the individual using probabilities, and these are not certain for you. You will never know which group you are in—whether you benefited from screening or were harmed by it. Screening can help narrow the odds, one way or another, but even relatively accurate screening seldom rules fatal risk out, or in, ever. Norm is right, sadly; we have no choice but to accept uncertainty even if that means more fear.

Prostate cancer is another classic example. The Prostate Specific Antigen (PSA) test, based on work by Richard Ablin in the 1970s, is used extensively in the United States for screening men without symptoms. Ablin now says that PSA testing is "a profit-driven public health disaster." Nevertheless, in the United Kingdom, the composer Andrew Lloyd Webber argued in the House of Lords that "all men over 50 should have the PSA test and GPs should be encouraged to encourage them to do so." How can such a remarkable difference in opinion arise?[8]

The problem is that, if you have no symptoms or other reasons to be at increased risk, this simple test can start you on a treatment regime ending in incontinence and impotence, as Lloyd Webber has honestly acknowledged.[9]

It is not difficult to find stories from people who claim their lives were saved by being screened, but it is, however, difficult to say what would have happened if they had not been screened. Maybe they would never have been hurt by whatever was found. It is extraordinary how much hidden disease there is, sitting around not causing any problems. When 2,000 healthy people with an average age of 63 had their brains scanned as part of a research project, 145 (7 percent) had had brain infarcts (strokes) that they had not noticed, while 31 (1.6 percent) had noncancerous brain tumors.

In people over 40 receiving diagnostic ultrasound, 14 percent of men and 11 percent of women were found to have gallstones, even though they had no symptoms.[10]

Studies at autopsy for deaths unrelated to any disease—in car accidents, say—reveal the astonishing amount of undetected cancer. In fact, there's around a 75% chance that either DS or MB (or both) has prostate cancer at the moment, given their ages, since "it is estimated from post-mortem data that around half of all men in their fifties have histological evidence of cancer in the prostate, which rises to 80% by age 80," says Cancer Research UK, although it goes on to point out that "only 1 in 26 men (3.8%) will die from this disease."[11]

Unfortunately, screening can't tell the difference between cancers that will turn nasty and those that will sit around minding their own business. Over the past 30 years, the United States has experienced a large jump in the number of prostate cancers diagnosed, but there has been only a moderate impact on the death rate, even with great improvements in treatment, suggesting the screening has not been of substantial benefit in reducing deaths.[12] But all this activity makes the "survival" rates look wonderful, as survival is measured from diagnosis, and so the start-time for "5-year survival" begins earlier, even if people are not living longer. By adding cases that would never have been noticed without screening, the survival statistics appear even better.

Working out the balance between the benefits and harms of screening is notoriously tricky, and the best way is through large trials in which thousands of people are randomly allocated to be offered screening or not. In the United States, 80,000 men were divided up in this way, and after 13 years the screened group had 12 percent more cancers diagnosed but no difference in deaths from prostate cancer. A European study of 182,000 men did find a 21 percent reduction in prostate-cancer deaths after 11 years in the group offered screening, corresponding to a reduction of 1 death per 1,000 men, so that 1,055 men would need to be offered screening and 37 additional cases treated to prevent 1 death from prostate cancer. But there was no effect on deaths from all causes.[13]

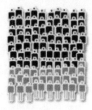

david spiegelhalter
31.3 out of 100
men of European ethnicity who share david spiegelhalter's genotype will develop Type 2 Diabetes between the ages of 20 and 79.

Average
25.7 out of 100
men of European ethnicity will develop Type 2 Diabetes between the ages of 20 and 79.

FIGURE 32. DS's risk of Type 2 Diabetes based on some genetic markers.

It is very natural for those who have endured diagnosis and treatment for cancer to attribute their survival to the test that first alerted them to their illness. So it is a potentially upsetting message that, of survivors of cancer found at screening, 90 percent would be alive had they not gone to screening in the first place.[14] People's innate storytelling habit is to connect one event with a preceding one: first this (screening), then that (treatment), leading happily ever after to survival. Not necessarily.

But if not in screening, can we find certainty in our genes? To check this, DS spat into a plastic tube and sent it across to the United States, so that a company called 23andMe could check (screen) some markers in his DNA and tell him whether he could blame his ancestors for his prospects.[15]

But they only provided lots of information about all the horrible things that he *might* develop, including telling him he had an increased lifetime risk of type 2 diabetes, which they did using the graphic in Figure 32. Will he end up being a figure in gray or black?

Well, he has already lived to 59 and has not gotten diabetes yet, so it's looking hopeful he might be one of the gray ones. These are just assessments of the risks using a few items of information about his genetic makeup that could have been obtained when DS was a baby. Now things have changed, and so the odds have, too.

To find out whether he had a genetic risk factor for Alzheimer's disease, DS had to tick a box agreeing that he really, *really* wanted to know this information before it would be unlocked. Which he did. And the answer was? He isn't telling.

So is this the future? And how much anxiety, investigation, and unnecessary medical care will it bring? Because if it's a future that chases reassurance, it might in a vast number of cases simply remind us, perhaps painfully, how much we will never know.

MONEY

Once more, the old memory haunted Norm: of being by the sea with his little brother, digging a hole in the sand just like they did on that endless beach at Skegness. He liked it outside, liked the sea and the air. It was the time Dad bought them those red metal spades with wooden shafts and handles. The spades were ace, they said, ace. He gave them money for ice creams, too. Norm felt the coins in the pocket of his shorts. He liked that: the roundness, the promise.

The sand was firm and just wet enough to stay up as they carved the sides of the hole, steep and square. Their arms dug, turned, and tipped. They liked how the sand gave way to the spade, how it cut in slabs. Then Norm's brother began another hole. Now there were two holes with a wall between them two feet wide. They cut foot-holes in the sides to climb out.

"A tunnel," said the little brother.

"Yeah, a tunnel," said Norm, "sort of."

Carving slices of sand, they joined their two holes with a tunnel. They widened the tunnel and widened it some more, slice by slice, with their red spades, squatting and cutting, watching for cracks in case it was going to fall. And when it was done they were glad and had two holes nearly as deep as the boys were tall. There was a bridge of solid sand, the tunnel perhaps two feet high.

"Crikey," said Dad, who came over with his trousers rolled up. "Can you get through?"

Norm's brother jumped in with a sandy thud, twisted onto his hands and knees. He was little. He squeezed through the tunnel and up the other side.

"Easy," he said.

"You'll be all right," said Dad to Norm. "I'll watch."

Norm jumped down. It was gray and a bit wet at the bottom of the hole. He hadn't noticed before, when he was digging. He peered through to the other side and saw that it was gray and wet there, too. He knelt down. The tunnel was dark and the roof low. He wanted to be quick like his brother was quick, and he dipped his head and shoulders and crawled, but his bum caught the bridge and for a moment he was stuck and then he backed up fast.

"Can't do it."

"Go lower."

Lower would be not flat, but not strong and quick like it was on hands and knees. It would be in between, and that was awkward and bad. It was bad to move his face and stomach closer to the sand, with so much sand on top. He dipped down, lower, and crawled, and bumped, and so he went lower again, deep in the smell of wet sand, further from the light, his hips low and his arms bent, his lips gritty and his face close.

And then his head was through. But his body was still in, with the thick block of sand over him, and he was staring straight at the sand wall of the side of the hole and the turn on this side was awkward, and he bumped the walls, bumped the roof, and he was stuck again and needed to twist, and puffing, something tight, he . . . around . . . that way . . . and was out, and up . . . and felt something go inside, and cried. So they jumped on it until it fell. Then he felt in his pocket, and it was empty.

Back in the present, Norm put on his robe and went downstairs to check the balance in his savings account and look again at his pension statement. Silly. He knew what it said. Silly, Norm, to feel so put out, so vulnerable, again, still.

· ·

AS IMAGES GO, penniless and in a hole isn't subtle. But then neither is the kind of fear that takes you by the throat. Or the strange way that weird associations form in the mind, and then resist our every effort to talk ourselves out of them. Nightmares and phobias don't go away because we tell ourselves not to be silly.

Norm is old. He's seen some life. He has experience. But his anxiety here goes back a long way, and for him no amount of wisdom or calculation has cured it. This state of mind is sometimes said to follow the doctrine of the searing memory: every judgment dictated by a deep mental scar. When you can still taste the panic in your mouth like sand, what chance the objective calculation of abstract risk?

For Norm, who trusts in logic, this hurts: once again, his mind won't obey his own orders, all because of one tyrannical moment years ago. Out of proportion? It makes no odds. For decades he has been telling himself to grow up and be reasonable, but little by little Norm has also been learning what it is to be human.

Phobia is an extreme example of the availability bias that we met in Chapter 4 (about how we change our minds about data according to how it's presented or framed). Availability bias, as we saw, means whatever comes easiest to mind. Everyone is affected by availability bias, although we've afflicted poor Norm more than most by giving him a phobia. It sticks in his mind. It comes to mind often. He can't help it. Sorry, Norm.

Daniel Kahneman argues that there is evidence people can fight availability biases if they are encouraged to "think like a statistician" and work out what it is that might be shaping their opinions. You can do this, he says, by asking questions such as: "Is our belief that thefts by teenagers are a major problem due to a few recent instances in our neighbourhood?" Or: "Could it be that I feel no need to get a flu shot because none of my acquaintances got flu last year?"

The phobia haunts him now because he feels his last years are especially vulnerable. He feels afraid of being penniless and powerless. It's not much of a sunset. But is it true? Or, if he's typical, will Norm burn the kids' inheritance on cruises for seniors, flush with an index-linked pension and fat on housing equity withdrawal?

What's certainly true is that people are living longer (see more on this in Chapter 26). That should be cause for celebration, except that so many seem to fear they won't be able to manage, with a future of misery and poverty. To others, the generations now reaching retirement have had it all—and seem hell-bent on blowing every cent. Whatever the truth, instead of the blessing of a long life, we lament the risks and burdens of age—either the burden of enduring it or the burden of paying through the nose so that others can soak up the sun.

So who is right, and which image of retirement is true: the risky, phobia-inspired nightmare of poverty and vulnerability, or the image of retirement as a twenty-year party?

Both, to some extent. There are pensioners who have done well, and there are plenty who have not. Since this chapter is about financial risk and insecurity in retirement and old age, we will concentrate mostly on the have-nots.

Historically, old age has not been a time of plenty. It is estimated that in 1900 about 5 percent of the elderly overall and about 30 percent of the over-seventies in Britain were in the workhouse, a grim, punitive institution still operating under the Poor Law of 1834.[1] Lives inside were often harsh by design, to deter the able-bodied, although workhouses did at least provide medical care.

But they were harsh for the infirm, too, and it's less well known that they became a common destination for the old. Still more elderly people relied on payments known as "outdoor relief"—small supplements to scraps of work, charity, or family help, so that with luck they could remain outside the walls of the "pauper bastilles."

Most elderly people relied on others to some extent. Many were well cared for, but a big minority were not, and "the extreme smallness of their means" was noted by social reformers, such as Charles Booth and Seebohm Rowntree. Only after the appointment of a royal commission in 1905 was it suggested that deterrent workhouses be reserved for "incorrigibles such as drunkards, idlers and tramps." George Orwell wrote of standing in the queue for "the Spike" with a tramp, "a doubled-up, toothless mummy of seventy-five."[2]

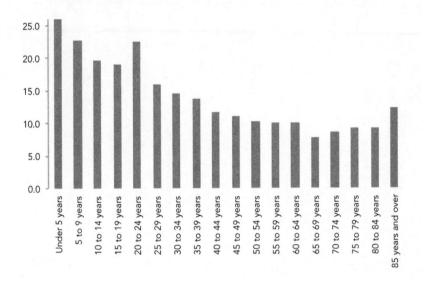

FIGURE 33. Percent below poverty in the United States, by age, 2010.

With the development of state pensions in the United Kingdom in the twentieth century, retired life steadily improved, although women were still disadvantaged, often receiving pensions only through their status as wives or widows. Even so, retired life slowly became less grim. Today, some say retirement has become a feather-bedded indulgence, with "gold-plated" pensions and voluntary early retirement at the top. Recently, some have argued that it's about to get worse again and that the next generation of seniors faces a meaner future.

There are two big fears here: first, not having enough to get by for so long as you can be independent; second, being cleaned out by the cost of care. But, as we've often seen, fear and data don't always line up. So what is the data?

According to the US Census Bureau, those aged 65 to 84 are less likely to be in poverty than members of any other age group. Above 85, there's a bit more poverty, but still less than for people in the prime working years of those aged 30 to 39. And this is despite the climbing health-care costs associated with old age. According to

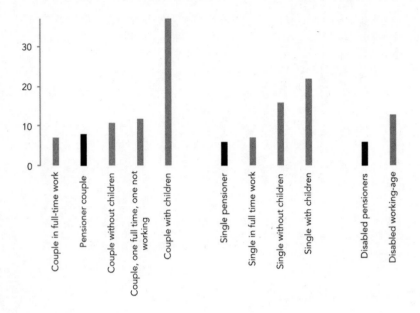

FIGURE 34. Percentage of different types of UK households "in poverty."

the Employee Benefit Research Institute (EBRI), over the past three decades, as poverty rates increased for younger age groups, they fell for older Americans, although in the five years to 2009, recession pushed them up again. One more striking fact: elderly American women are twice as likely as men to be poor.[3]

In the United Kingdom, too, current retirees have reached the surprising point where, as a group, they are less likely to be poor, on average, than almost any other segment of the population (see Figure 34), and their rate of poverty has fallen sharply, by about 45 percent since 1999. Poverty in old age is still a fact of life for nearly 2 million people in the United Kingdom, but the situation has improved rapidly.[4*]

---

* The standard way of measuring income in the United Kingdom is by household. Since different households have different numbers of people in them, the figures are adjusted to take account of this. So it is assumed that a single person living

In fact, the poorest Americans, like the poorest Brits, often do better after retirement than before. EBRI says that in every survey year, poverty rates in the 65- to 74-year-old age group are lower than in the 50- to 64-year-old age group, which may be because Social Security payments have kicked in. Though, as the elderly age further, a few more slip into poverty.

So the image of the impoverished senior citizen is misleading. Not because some aren't struggling—they are—but because this is not mainly a characteristic of the old. It is a characteristic of the poor—of all ages—and it tends to be less true of the old poor than the younger poor. This makes an important point: most people who are classified as poor in the year in which they retire were probably poor before. Retirement probably didn't cause the poverty.

Higher up the income scale, people do tend to take more of a hit. Even so, overall and in the long term, life in retirement has improved dramatically, especially for the worst-off. Seniors over the long run have become steadily less likely to be classified as poor. In fact, real median incomes for seniors in 2012 (i.e., not dragged up by the influence of the super-rich) were close to their all-time peak, bettered in only one previous year—pretty remarkable compared with others, who've seen steep declines.[5]

Elderly Americans can still face steep health-care costs, even when enrolled in Medicare. But they might be surprised to discover that in the United Kingdom, considered by its citizens to be the cradle of the welfare state, there is still one cost that can be ruinous. Not health care, strictly speaking, which is free at the point of use, but nursing care outside of the hospital—to some a weird distinction. Although you can insure yourself in the United Kingdom against unemployment or illness, you can insure against your house burning down or accidents at home, on the road, or abroad, you can insure your pets, and, in financial markets, you can even buy

---

alone needs 67 percent of the income of a couple to achieve the same standard of living. Similarly, people with children need more than a couple. By this means, all households can be compared on the same scale, approximately.

end-of-the-world insurance, you cannot insure against the potential costs in old age of what the United Kingdom calls "social care," and the state provides financial help for this only when you're almost wiped out.

In 2012 this was just about the only potentially expensive life-time risk that could practically empty the coffers with no way to stop it. There was simply a messy and often inconsistent system that could punish the unlucky with the loss of all their assets, including the house. Even at that price, the quality of care varied hugely.

If you can't insure against a risk, you have no choice but to take the cost on the chin if the worst happens. That can screw up life for everyone, even those who turn out not to need care, because the problem is that you never know if it will be you.

About 1 in 10 Brits can expect costs like these of over £100,000 in their final years, and there is no knowing who, no way of pre-dicting in advance what the costs might be for any individual.[6] For those who have little or nothing, the state will pay. For the rest, this ruinous tail-end risk is frightening, and it frightens insurers out of the market altogether. At the time of writing, the UK government is considering proposals to increase state help for those with extreme costs in order that a market might develop for the rest, but it is still not clear at what level the help will kick in, and these reforms are not expected to be implemented until 2017.

For this reason, although in most other respects old age is more financially secure than ever, Norm is right: life after retirement is still a financial lottery in which he could lose almost everything.

26

THE END

The asteroid missed. The odds that it would strike earth had lengthened, as they tend to do following initial discovery, once the trajectory is refined. Even with years to go, it was clear the Apocalypse was postponed, again.

Stargazers watched the fly-by in fascination, with a quiet thrill at the thought that this was the one that could have been. "What if? . . .," they said.

Norm stayed inside. And S043 passed quietly about halfway between the Earth's atmosphere and the moon, then sailed into oblivion. It had all been a false positive. A non-event. House prices rose slightly.

In the evening he stood straight and proud in his pajamas, a stub of Aquafresh toothpaste on the toothbrush that hung by his side. He noticed a slouching Norm reflected in a dark window, as if his appearance were news to him. He didn't bother with all that anymore.

A minute passed. The stub of toothpaste began to peel off. The pipes of the central heating warmed and stretched.

"Any pike out there?" Norm said to the reflection.

No answer. He'd asked before but was less sure nowadays that he wanted one. Did it matter—the answer? What if someone told him, and then told him what to do about it, laid out all the answers? Norm sighed. No need for choices if all the answers are given.

He smiled. Then he looked at Norm with narrowed eyes.

"I see through you," he said.

Paler and paler these days, it was almost true. Growing old lent his skin a soft transparency that was not white, nor gray, nor grayish, nor reddish, nor brownish. Tupperware skin, Mrs. N said. Tupperware hair, too. People ignored him. They said they couldn't make out what he was saying.

"Everything all right?" she called from the other room.

"Fine," he said.

Norm held his own gaze, fixed as a Rembrandt. All his life he had tried to reduce fear to probability. Strangely, he felt no fear now.

"Probability? . . . Doesn't exist," he said, just to be rude to the man in the window, who was rude back, sharing one feeble voice. He stared harder, his eyesight so poor these days he could hardly make out . . .

"The average of what?" he mumbled . . . what with the poor light, feeling almost weightless himself—as the toothbrush dropped.

No pike, then, not for him. He reached down with one hand like the boy who stood—was it on the bank, or was it a dream?—limbs frail and light, and lighter, as he reached, or thought he reached, and felt nothing. Curiously, Norm had disappeared.

Mrs. N woke the next morning and rolled over. As she made tea downstairs, Kelvin called about wanting a lift. "Have you seen Norm?" she asked.

"Who?" said Kelvin. He was old. But perhaps he was right.

"Never mind."

Later, Mrs. N noticed Norm's pajamas, crumpled in the bathroom, and a toothbrush on the floor.

Later that day, coming out of the bookie's in his wheelchair, Kelvin had a heart attack—all those cigarettes and burgers, probably—and went out suddenly. It was not abnormal for a man of his age and habits, which would have irritated him no end. There's an average even for unusual people: the unusual-people average, and Kelvin, unknowingly, was typically unusual.

Years later, Prudence had her last coherent thought about Norm as she dribbled into her All-Bran and hoped that, wherever he was, he was safe, but she felt horribly confused by the strangeness of it all, and it wasn't long before she had forgotten him again. Having looked after herself so well for so long, her body lasted until her mind gave up. She drifted on for five more years or so, happily unaware at last, loved by her family until the end.

On the basis that at least one unusual thing happens to everyone, on average, even to the average, it could be argued that Norm's strange disappearance and presumed death were unsurprising. Except that for some, the unusual thing that happens is that nothing much happens at all.

. . . .

DID NORM HAVE TO DISAPPEAR? We'll answer that in the next, and last, chapter. Though they all had to go somehow. In that respect, death isn't a risk at all. It has a probability of 1. All that's risky about it is that the reaper turns up sooner than you think right or proper. For some of us, the right moment can never arrive. For Kelvin death seemed like one more rule, so of course he'd have wanted to rebel. For Pru, when the risk really was creeping right up behind her, just for a change she couldn't care less. But the timing was, for all of them, in its way normal.

And normal is? . . .

The 90th Psalm in the King James Version of the Bible declares that "the days of our years are three-score years and ten," although up until recently you had to be fairly lucky to reach this use-by date. No English monarch survived until 70 until George III, in 1820, although they did have to cope with drafty castles, bad sanitation, and occasional violence. Some historical figures managed it: Augustus Caesar conquered 75, Michelangelo hammered on to the amazing age of 88.

The number 70 was important for another reason: since the ancient Greeks there had been a superstition about the risks of ages that were multiples of 7. In particular, the "climacteric years" of 49 and 63 were thought positively dangerous.

Age. Curt.	Per-fons.	Age. Curt.	Per-fons	Age. Curt.	Per-fons	Age. Curt.	fons	Age. Curt.	Per-fons	Age. Curt.	Per-fons	Age.	Perfons.
1	1000	8	680	15	628	22	585	29	539	36	481	7	5547
2	855	9	670	16	622	23	579	30	531	37	472	14	4584
3	798	10	661	17	616	24	573	31	523	38	463	21	4270
4	760	11	653	18	610	25	567	32	515	39	454	28	3964
5	732	12	646	19	604	26	560	33	507	40	445	35	3604
6	710	13	640	20	598	27	553	34	499	41	436	42	3178
7	692	14	634	21	592	28	546	35	490	42	427	49	2709
Age Curt.	Per-fons.	Age. Curt.	Per-fons	Age. Curt.	Per-fons	ge Curt.	Per-fons	Age. Curt.	Per-fons	Age. Crit.	Per-fons	56	2194
43	417	50	346	57	272	64	202	71	131	78	58	63	1694
44	407	51	335	58	262	65	192	72	120	79	49	70	1204
45	397	52	324	59	252	65	182	73	109	80	41	77	692
46	387	53	313	60	242	67	172	74	98	81	34	84	253
47	377	54	302	61	232	68	162	75	88	82	28	100	107
48	367	55	292	62	222	69	152	76	78	83	23		34000
49	357	56	282	63	212	70	142	77	68	84	20		Sum Total.

FIGURE 35. Halley's original calculations for what we would expect to happen to 1,000 people starting in their first year.

Partly in order to combat this belief, in 1689 a priest in Breslau in Silesia (now Wroclaw in Poland) collected the ages at which people died. The data eventually found their way to Edmond Halley in England, who took time off from discovering comets to construct the first serious life table in 1693, which uses estimates of the annual risk of dying to work out the chances of living to any age.

He found no evidence of any increased risk at age 49 or age 63, so he neatly demolished the idea of climacteric years. His comet also duly turned up, as predicted, in 1758, when Halley would have been 101, if the force of mortality had not finally got the better of him. Halley's life tables only went up to 84: he estimated there was a 2 percent chance of reaching this age, and just to prove it, he died when he was 85, thus by some margin outliving Bill Haley, also of Comet fame, who rocked around the clock only until 55.

"Force of mortality" was, in fact, a technical term for the annual risk of death, now known as the "hazard" of mortality. The rough shape of the hazard curve shown in Figure 36 seems to be a constant throughout history: a high hazard just after birth declines to a

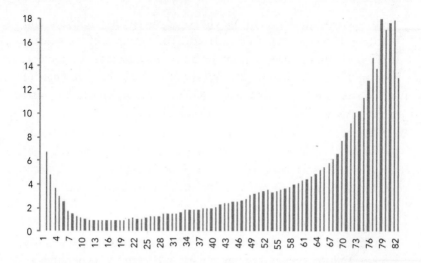

FIGURE 36. The annual risk of death—the "force of mortality"—derived from Halley's data for the 1680s.

point of greatest safety and then starts increasing again, although the precise pattern in childhood and young adulthood depends on the contemporary state of infectious diseases, war, and, nowadays, recklessness with drink, drugs, and driving.

In 1825, Benjamin Gompertz, barred from university study for being Jewish, formulated his "law of mortality," which says that after the mid-20s the annual risk of dying increases at a constant rate. This is extraordinarily accurate, and every year between the ages of 25 and 80 the risk of dying has a relative increase of around 9 percent, as already mentioned in Chapter 17. Gompertz did his best to fight against his own law but finally died when he was 86.

Longevity is usually summarized by life-expectancy—the average length of life—but averages can be misleading, as we've seen so often. Life-expectancy is strongly influenced if there are a lot of deaths in childhood, in which case the survivors may live to a good age, but the average will still be low.

In 1958, when he wrote "When I'm 64," Paul McCartney was only 16, an age when 64 seems old—and particularly in an era when people thought retirement meant putting on your slippers and waiting for death. But even the chance of surviving childhood and reaching 16 has changed dramatically. For example, if we consult that wonderful resource the Human Mortality Database, we find that back in 1841, a frightening 31 percent of children born in England and Wales died before they were 16.[1] But if you survived, there was nearly a 50 percent chance of reaching 64.

By the time the Beatles recorded "When I'm 64" for *Sgt. Pepper's Lonely Hearts Club Band* in 1966, only 2.5 percent of children died before 16, and surviving girls had an 85 percent chance of reaching 64; for boys it was 74 percent, the difference partly reflecting the unhealthy lifestyles of so many men.

By 2009, less than 1 percent of children died before 16, and after that the chance of reaching 64 had risen to 92 percent for women and 87 percent for men, giving an overall life-expectancy of 82 for women and 78 for men.

There have been similar gains in the United States, where average national life-expectancy at birth was about 69 in 1970. In 2011 it was nearly 76.3 for men and 81.1 for women, lower than in the United Kingdom, but still a huge increase. But these figures depend crucially on where you live, which is often a function of where you can afford to live, which in turn is often associated with racial background. By state they range from an average of nearly 83 in Hawaii to 75 in Mississippi. By county, the differences would be bigger still.[2]

In the United Kingdom, if you are lucky enough to be a resident of one of the richest places in Britain, Kensington and Chelsea in London, then a mostly well-heeled lifetime of tea and crumpets, or more likely these days, gazpacho and twice-cooked pork belly with an onion and apple velouté, is expected to last until 90 for females and 85 for males, whereas in Glasgow City in Scotland rather different consumption and material lifestyle is associated with an extraordinary 12 fewer years for females and 13 fewer for males.[3]

Although it's a measure of the speed at which people are putting on extra years of longevity that today's disadvantaged Glaswegian enjoys the same expected survival as the average man in England did in 1983, just as life-expectancy in Mississippi is about where the whole United States stood, on average, in 1990.

In fact, life-expectancy in the United States has been racing up for generations—by more than 2 months a year, on average, over the past 50 years, faster for men, slower for women. That is as if, after using up 48 MicroLives just slogging through a day, you are given about 10 of them back again, every day, by those nice people who build drains, give us injections, stop us from smoking, sell us low-fat milk, and treat us in the hospital.

But this is nothing to the extraordinary changes that have occurred elsewhere. In 1970, people in Vietnam had a life-expectancy of 48. It is 75 now, a transition that took England and Wales more than twice as long to make, from 1894 to 1986.

Behind the cold columns of numbers in historical tables lie powerful events: apart from the world wars, Napoleon's march on Moscow (which killed 400,000 men) temporarily reduced life-expectancy to 23, the influenza epidemics of 1918–1919 took 10 years off of the life-expectancy for females in France, while AIDS has meant that life-expectancy in South Africa declined from 63 in 1990 to 54 in 2010. In 1901, life-expectancy for black males in the United States was 32, with 43 percent dying before the age of 20, while for white males the corresponding figures were 48 and 24 percent. After 100 years, the gap in life-expectancy was still there, although it had narrowed from 16 years to 5.[4]

All the "life-expectancies" quoted above are based on accumulating the current annual risks of dying—the force of mortality at the moment—and do not take into account any future progress. If we're going to say something about the prospects for children born now, or yet to be born, we have to make some assumptions about what is going to happen to human health and life span. The "principal projections" for England and Wales estimate that males born now, allowing for projected future improvements in health, can, on average, expect to live until 90, and for females, 94. Thirty-two percent of males and 39 percent of females are expected to reach 100 and get a letter from the Queen, or whoever has the job in the early twenty-second century.

Babies born in 2050 are projected to live until an average of 97 if male and 99 if female, meaning they will die around 2150. Each generation has an ever-longer reach into the future.[5]

But these projections are, rather understandably, deeply controversial. Is there some inbuilt aging process, and can it be reversed? Or is there some natural ceiling that we are going to bang our wrinkled, bald heads against? The arguments are fierce. Put a group of specialists on aging together in a room, and you will be lucky if there are any survivors. People have claimed for years that a ceiling is being reached, and yet life-expectancy just keeps going steadily up. Individual extremes catch attention: Jeanne Calment was born in 1875, soon after the Franco-Prussian War, but kept soldiering on until 1997, when she died at 122, and there are now more verified 115-year-olds than you can shake a walking-stick at.

One thing is certain for children born in the future—there are going to be a lot of old people around for them to look after, as those already plodding through their middle age are not going to go away. The United Nations estimates that the proportion of people over 60 will double between 2007 and 2050, as people will live longer, and lower fertility rates mean fewer young people. There will be 2 billion people over 60 in the world by 2050, and around 400 million people over 80.[6]

But what sort of state are all these old people going to be in? After the bit about three-score years and ten, Psalm 90 continues with "and if by reason of strength they be fourscore years, yet is their strength labour and sorrow; for it is soon cut off, and we fly away," which doesn't exactly paint an encouraging picture of aging. Are we making all this effort to live longer just so that we can spend even more time sitting around the edge of a room, television blaring, struggling to grasp what reluctant visitors are shouting at us?

But are there more positive indications of health in old age? Well, there are in the United Kingdom, but less so in the United States. It's not just life-expectancy in the United Kingdom that's growing, but *healthy* life-expectancy—the number of years spent in good health. Britain's Office for National Statistics says that healthy life-expectancy is currently rising even faster in the United Kingdom than life-expectancy as a whole, a process described as the "compression of morbidity." But on limited data, although healthy life-expectancy in the United States is also rising, it looks as if it's rising more slowly than life-expectancy as a whole. That is, a smaller proportion of American life spans are years of good health. This might be because people judge the quality of their own health more severely than they used to, rather than because it has really changed, or maybe it's that Brits are more easily satisfied. Either way, Americans feel they have more to complain about than they used to.[7]

What's more, about 35 percent of the people 75 and over in the United States live alone, and loneliness is associated with failing mental abilities, which is probably feared even more than physical disability. This is already a big issue, and it won't become any easier. According to the Centers for Disease Control and Prevention, between 25 and 50 percent of Americans aged 85 and up exhibit signs of Alzheimer's disease. Up to 5.3 million Americans currently have it, a number expected to more than double by 2050 due to the aging of the population. Alzheimer's disease is the sixth leading cause of death in the United States.[8]

Figures for the United Kingdom are of similar proportions. In 2005 it was estimated that 700,000 people were living with dementia in the United Kingdom, and this figure is likely to rise to 1.7 million by 2051. Nearly all of these people who will have the disease then are already alive now. Perhaps you are one of them.[9]

In 1958, Paul McCartney saw 64 as near the end of life, whereas baby boomers who were born just after World War II and have now reached this age tend to have a reasonable lifestyle and only see themselves as "middle-aged," with much more before them than a precipitate decline. In 2008, UK men at age 65 had around 17 years of life-expectancy left, 10 of which would be considered "healthy," while women had more than 20 years left, on average, 11 of which would be "healthy"—"healthy" means that people consider their health as "good" or "very good" on a 5-point scale, which is quite a stringent criterion.[10]

So what should these people be doing? Oddly, the proximity of death can make risk seem less risky. That sounds strange. But if you looked simply at relative risk probabilities, you could almost conclude that the old should be out there playing with fire. Here's how that surprising calculation works.

Recently, DS was asked to do a tandem skydive for a TV program, and, of course, he checked the data. Skydiving has clocked in at an average of around 10 MicroMorts over the past 20 years in the UK, but he reckoned tandem would be a bit safer (there has been only 1 death in 340,000 tandem jumps recorded by the British Parachute Association[11]). So he reckoned around 7 MMs for the plane ride and jump.

For a strapping young British lad of 18 contemplating the same jump (average total annual risk of death from all causes 530 MMs), this is equal to around 5 days of average background risk, whereas for a semi-wasted man DS's age (all-cause annual risk 7,000 MMs), it is only around 9 hours' worth. So, from a relative perspective, it makes more sense for an old codger to hurl himself out of a plane, charge around on a Harley-Davidson, and play chicken on highways than it does for some cocky youth with his life ahead of him. But try telling that to the youth: they never listen.

27

JUDGMENT DAY

We—that is, the authors—have a problem: we like numbers and we think they matter, but we like our characters, too. Call this vain—we invented them, after all—but Norm, Prudence, and the Kevlins are mostly all right, we think. More than that, although we recognize the patterns in numbers as well as anyone, we can't bring ourselves to say that any of our characters are irrational when they choose to ignore the numbers and go their own way.

It has become fashionable to blame people for their cognitive biases or failures of reasoning, but we think that much of this so-called irrationality about danger is really a result of the way the information is framed, or the sheer complexity of the decision, and that their reasoning is usually reasonable, for all their human ways.

It's not reasoning that we would necessarily share, but nor is it reasoning we can easily dismiss. You can call Norm a fool if you like, Kelvin an offensive slug, and Prudence a pain in the arse—though we think they deserve more understanding than that—but given what they want from life, we don't know how to prove that any of them are actually *wrong* about the choices they make in the face of danger. We don't know how to use data to tell them how to live. Even if planes feel more dangerous to travel in than cars, it's not much use to point out to any individual that the probabilities of fatality are the other way round, since what they mean by danger is seldom as simple as a mortality rate. It is often, as we've said, not even about danger.

They may be odd, but they are not stupid, and they know what they care about. They live in an uncertain world, where risks can change and no one knows which side of the odds they'll face. Their hopes and fears are not like the unfounded fears of monsters under the bed; they are fears in a real and usually messy human context.

This is not to turn all aspects of danger into relativism. Our view is that planes really are safer than cars, on average. We simply argue that a measure of what people call risk must be a matter of personal value and personal framing. The objective numbers can't be separated from subjective perception. A risk stated as 1 in 400 can't help but focus attention on the 1. To say that the risk is the same if we describe it as a chance of 399 in 400 of nothing happening might be mathematically correct, but when people react to these different frames differently, that does not, in our view, demonstrate their irrationality. It demonstrates the importance both of the numbers and of perspective. Think of it like the view of the countryside from the city, compared with the view of the city from the countryside. Both country and city exist in the same proportions wherever you stand, but that doesn't mean that where you stand is irrelevant. The view matters. So do the proportions. Risk does not exist independently of the way that people see it. Nor can the way they see it float free of the numbers.

We could go further. We looked at a lot of numbers to write this book—and we mean *a lot*—and we have a geeky streak ourselves. Possibly you noticed. Maybe that leads you to expect that our private attitude toward risk will be on reason's side. And on the whole it is. But while we think the numbers are vital, we are not piano keys either, and we share a deep sense of the uncertainties around data, statistics, and evidence. Whether you think that chance really exists, or that your fate was cast in the Big Bang, or that it's all fixed by a deity or three sisters spinning, measuring, and cutting your fate, we all still have to deal with not knowing quite what will happen. We think there's more uncertainty than you'd think from the way people throw numbers around.

We think this especially because when you try to grab hold of probability it somehow slips through your fingers. It's hard to say what it really is for an individual. It's hard, too, to say how the average affects any individual.

Of course, some things are more likely than others: the chance of getting wet if it rains compared with being—alone of 7 billion people on the planet—the one to be struck by an asteroid. Ignore these differences and we would worry about you. It shouldn't have taken until the 1650s, when Blaise Pascal and Pierre de Fermat began a correspondence about dice, for people to get around to putting numerical probabilities on events. But although in some ways it is a simple idea, it has caused headaches ever since.

As an example, on April 14, 2012, DS saw a tip in a newspaper and put £2 on Cappa Bleu to win Britain's Grand National horse race. The odds were 16 to 1, which meant a profit of £32 for a win, but could also be interpreted as meaning that the bookmaker thought there was a "chance" of around 6 percent that Cappa Bleu would win.

Previously, DS had been to see his general practitioner, who measured his blood pressure and cholesterol, tapped at his computer, and duly announced a 12 percent chance of a heart attack or stroke in the next ten years. Disturbing . . . until the GP also said that this was less than average for a man of the professor's age, who then somewhat irrationally cheered up at this change of framing. After a strong finish, Cappa Bleu came in fourth.

But what do all these probabilities mean? Philosophers and statisticians have argued for centuries and are far from agreement. Into this sometimes savage controversy what can we do in a short space but charge, prejudices blazing? Here are ours, which mirror how we feel about Norm, Kelvin, and Prudence.

Traditionally, probabilities have been based on known physical properties and pure reasoning—for example, a coin has two sides, so a coin toss has a 1 in 2 chance of coming down heads; throwing a die has a 1 in 6 chance of a six; dealing one card from a (properly) shuffled standard deck of cards has a 1 in 52 chance of turning up the ace of spades. But this only works if we start with some idea of "equally likely events," in the sense that all cards are somehow equally likely to turn up. But that requires us to say what "likely" means, so we're back where we started. (And then, of course, in real life someone might cheat.)

Another idea of probability is to say that it is how often something happens when a similar situation is repeated a vast number of times, such as the proportion of people who reach 100 years of age. But apart from very special set-ups such as flipping coins, these exactly similar situations just don't occur: there is only one you to reach 100, or not, only one Grand National race that DS could bet on in 2012, and only one DS to have a heart attack, or not. The idea that we all conform to the frequencies of the dead takes most of the life out of life. What was is not necessarily a good guide to what might be, let alone what will be.

As a way of dealing with the one-offs that we all like to think we are (yes, we are all individuals) some philosophers offer the idea of an intrinsic tendency for an event to occur, so that all the vastly complex aspects of DS's current and future existence combine to give some sort of "propensity" for him to have a heart attack or stroke in the next ten years, and the doctor's "12 percent chance" is an estimate of this. The idea of Cappa Bleu having some true underlying propensity to win is attractive but does not seem useful or provable.

We reject all these explanations and take a very pragmatic stance—that this "12 percent chance of a heart attack" is not really DS's risk, and not even an estimate of some propensity of it. It is based on a few items of limited information, and should be treated like the "probability" of Cappa Bleu winning—reasonable betting odds given current information. No more and no less.

Treating probability like a bet seems a cop-out, but it has powerful implications. It means that any number we claim for a "probability" is constructed by us based on what we know. It is necessarily a judgment and does not exist as a property of the outside world. Risk, in this sense, is a measure of what we don't and can't know as much as a measure of what we can.

All of which forms part of a rather startling conclusion: that independent, objective probability, as Norm says at the last, doesn't exist.*

Nor, as we say, does the average person exist to whom the average risk is supposed to apply. The average is an abstraction. The reality is variation. Poor Norm, he was a man in search of data, but perhaps the data were really in search of him. Was he ever really there? He had to disappear once he finally stopped believing in the norm.

In some ways probability is a bit like the certainty that there will be people, the story with which we began, but a certainty that tells you nothing about who. In other words, it tells you only a tiny part about you. The probability that there would be at least one person called Norm—high—is only a tiny part of the story about how this particular—infinitely improbable—Norm came into being.

So in practical terms, for the events of life in general, when we say a certain activity is dangerous and quote its risk as so many Micro-Morts, these numbers should be considered only as reasonable betting odds given what we know. As soon as we know more—maybe the age of the person about to try base-jumping and whether they are sober, or how many pike there are in the reservoir, if they're nearby, how fast they swim, when they last ate, whether the particular one that hasn't eaten lately still feels hungry and is near enough and fast enough and is the sort to recognize the edible potential of human flesh, even disguised by sagging underpants, and is a hard-nosed, vicious-enough, brave-enough sod to go for a lad's tackle—the risk changes, suggesting that the potential degree of refinement is often infinite. And this can just as easily be applied to things that have happened that we don't know about yet, such as giving the odds that Jack the Ripper was really the Duke of Clarence. Or Queen Victoria, for that matter.

Nor can any single risk capture the full complexity of our feelings and judgments about nature or the economy—or let's go the

---

* We concede that at the subatomic level there may well be irreducible, unavoidable, and inevitable chance, which means that "determined probabilities," to use Stephen Hawking's expression, could be said to exist. But this does not seem particularly relevant to our judgment about who will win the Grand National.

Cause of death	Context	MicroMorts	Exposure	Source
"Non-natural causes" (excluding suicide)	US, 2010	1.3	Per day	(a)
Complications of pregnancy or birth	US, 2010	625	Per birth	(b)
Infant mortality (first year)	US, 2010	6,100	Per birth	(b)
Infant mortality	World, 2010	40,000	Per birth	(c)
Infant mortality	Sierra Leone, 2010	119,000	Per birth	(c)
Accidents – under 14	US, 2010	63	Per year	(a)
Drowning – under 5	US, 2010	25	Per year	(a)
Pedestrians – under 15	US, 2010	4	Per year	(a)
Murder/homicide	US, 2010	48	Per year	(d)
Murder/homicide – under 1	US, 2010	36	Per year	(d)
Murder/homicide – age 20–14	US, 2010	106	Per year	(d)
Anesthesia for non-emergency operation	US	5	Per operation	(f)
Giving birth	US, 2010	210	Per birth	(g)
Giving birth	World, 2010	2,100	Per birth	(g)
Coronary artery bypass graft	US, 2008	18,000	Per operation	(j)
Serving in Afghanistan	US forces, 2010	22	Per day	(k)
Flying in Bomber Command in Second World War	RAF, 1939–45	25,000	Per mission	(l)

FIGURE 37. MicroMorts per hazard.

Cause of death	Context	MicroMorts	Exposure	Source
Walking	UK, 2010	1	Per 27 miles	(m)
Cycling	UK, 2010	1	Per 28 miles	(m)
Riding a motorcycle	US , 2010	1	Per 4 miles	(m)
Driving	US, 2010	1	Per 240 miles	(m)
Train	US, 2010	1	Per 6,300 miles	(n)
Commercial aircraft	US, 1992–2011	1	Per 7,500 miles	(p)
Light aircraft ("general aviation")	US, 1992–2011	1	Per 12 miles	(p)
Scuba diving	UK, 1998–2009	5	Per dive	(q)
Hang-gliding	UK	8	Per jump	(r)
Rock climbing	UK	3	Per climb	(r)
Skydiving	US, 2008-12	7	Per jump	(s)
Running marathon	US, 1975–2004	7	Per run	(t)
Ecstasy/MDMA (mentioned on death certificate)	E+W, 2003–2007	1.7	Per week	(u)
Heroin (mentioned on death certificate)	US, 2012	25	Per day	(u)
Asteroid	World	1	Per lifetime	(v)
Coal mining	UK, 1911	1,190	Per year	
Commercial fishing	US 2012	1100	Per year	
Logging	US, 2012	1280	Per year	
All occupations	US, 2012	32	Per year	
All occupations	World	160	Per year	

FIGURE 37. (continued)

whole hyperbolic hog and add the meaning of life—and so none of the natural, economic, lifestyle, or other risks that we talk about can be explained except in the context of a vast swath of other values. Our psychological reactions can be both optimistic (Kelvin and sex) and pessimistic (Norm and pike), and who knows in any instance which will apply? Similarly, no calculation of probability *times* consequence can tell you what weight you should attach to the consequences—in the unlikely event that you know them all—a weight that can only be for you to decide. And if half the calculation of a risk is infinitely variable, what is the objective answer to that calculation?

Probability sounds sensible enough, but whenever you reach for a firm and meaningful definition, the concept loses shape—although it is a number, show us the scales or the stick that you can measure it with. Egyptians, Greeks, Babylonians, and others did amazing things with algebra, geometry, number theory, and much more, but they never even got started on probability, and the omission is telling. DS says he has spent many years trying to work out why people find probability intuitively difficult and confusing. He has concluded that it is because probability is intuitively difficult and confusing. MB adds that he has often reported people's communication of risk and found the communicators don't really know

Factor	Definition of daily exposure	Males over 35		Females over 35	
		Estimated change in life-expectancy (years)	MicroLives per day	Estimated change in life-expectancy (years)	MicroLives per day
Smoking	Smoking 15–24 cigarettes (a)	– 7.7	– 10	– 7.3	– 9
Alcohol	First drink (10 g alcohol) (b)	1.1	1	0.9	1
	Each subsequent drink (up to 6)	–0.7	– ½	–0.6	– ½
Obesity	BMI: per 5 kg above 22.5 (c)	– 2.5	– 3	– 2.4	– 3

FIGURE 38. Average MicroLives (1/2 hour of life-expectancy) lost or gained per day of exposure to a specified hazard ratio for all-cause mortality, averaged over life after 35.

		Males over 35		Females over 35	
	per 5kg above optimum weight for average height	− 0.8	− 1	− 0.9	− 1
Sedentary behavior	2 hours watching television (d)	−0.7	− 1	− 0.8	− 1
Red meat	One portion (85 g, 3 oz) (e)	−1.2	− 1	− 1.2	− 1
Fruit and vegetable intake	Five servings or more (blood vitamin C > 50 nmol/l) (f)	4.3	4	3.8	4
Coffee	2–3 cups (g)	1.1	1	0.9	1
Physical activity	First 20 minutes of moderate exercise (h)	2.2	2	1.9	2
	Subsequent 40 minutes of moderate exercise	0.7	1	0.5	½
Statins	Taking a statin (j)	1	1	0.8	1
Air pollution	Living in Mexico City rather than London (k)	0.6	− ½	0.6	−½
Gender	Being male rather than female (l)	− 3.7	− 4	−	−
Geography	Resident of Sweden vs. Russia (m)	− 14.1	− 21	− 7.6	− 9
Era	Living in 2010 vs. 1910 (m)	13.5	15	15.2	15
	Living in 2010 vs. 1980	7.5	8	5.2	5
Single dose of ionizing radiation	0.07 mSv (e.g., single transatlantic flight)	30 mins	−1	30 mins	−1

FIGURE 38. (continued)

what they are communicating. Just when anxious people most want clarity, they find muddle. There is a reason for that. It is a muddle. Norm, sadly, could never work that out.

The view that probability doesn't exist is unusual but moderately respectable.[1] It is also liberating, as it means that we are free to use a variety of metaphors and analogies when talking about risk, chance, or probability, free to look at it from multiple perspectives and accept that perspective matters.

For example, a 12 percent risk of a heart attack is often communicated as, "Out of 100 men like you, in 10 years we expect 12 to have a heart attack or stroke." But there are not 100 men like you, and the probability is not yours. A more gripping metaphor might be to say, "Of 100 ways that things may turn out for you over the next 10 years, in 12 of them you will have a heart attack or stroke."*

So, which of the 100 are you this time? And when Prudence wonders if she is the one in 100—and says "What if?"—Does she have a point? Or when Norm says she exaggerates, doesn't he have one, too? And when Kelvin says, "So what?" is he wrong?

Perhaps you have made your peace with the dual nature of risk, chance, and probability, and you are comfortable with its elusiveness from a personal point of view. But if Norm, Prudence, and Kelvin are not wrong to say what they say—each different, each to their own view of risk—then which is the right number for them? On the other hand, if you do happen to look up and see a falling piano. . . .

---

* The Bank of England uses a similar metaphor when communicating uncertainty about its forecasts with fan charts, in which the line describing predicted economic growth or inflation blurs out into a huge fan of graded possibilities, expressing what might be expected "if economic circumstances identical to today's were to prevail on 100 occasions." You can ask a computer to play out these futures multiple times, a technique known as a "Monte Carlo simulation," which started with the US project to build a hydrogen bomb. Similarly, there are "ensembles" for weather forecasting in which a number of different predictions are based on slightly perturbed assumptions about what is happening now, and chaos means that these small differences may result in wildly different predictions after a few days. Unfortunately, there is still a reluctance to talk in public about the chances of different weather patterns, although in the United States "possible paths" of hurricanes are shown on public news broadcasts.

ACKNOWLEDGMENTS

Andrew Franklin of Profile Books first suggested that we write together, an idea for which we're hugely grateful and which lacked only the detail of something to write about. For that detail, we owe thanks for two professional lifetimes of influences and opportunities and innumerable (even for us) friends and colleagues who helped us along the way. The Profile team has been a joy to work with, as ever. Jonny Pegg has been all you would want from an agent—in fact a double agent—a cogent critic and an affable, encouraging friend. T. J. Kelleher at Basic Books in the United States was likewise a dream of an editor—sharp and informed, enthusiastic and helpful, and those on his team were hugely impressive and conscientious. Andrew Dilnot gave invaluable advice at an early stage. Rich Knight and Chris Vince read some early material and said the idea wasn't unhinged, which helped. Katey Adderley and Caitlin Harris cheered MB along and were a mine of psychological insight about risk, while Joe Harris gave him his most important lesson in how to live with it. Edgar and Kieran supplied the outline of the story about penguins, years ago and from life. And we shamelessly plundered the thoughts about danger of any other friends foolish enough to talk about it. Kate Bull gave unlimited encouragement and advice, and even liked it, which was the main thing. Mike Pearson has provided endless support and inspiration. David Harding's generosity gave DS the opportunity to spend his time writing this sort of stuff. Thanks to them all.

NOTES

Note: All web addresses were live when last accessed on April, 8, 2013, unless noted otherwise.

INTRODUCTION

1. Steven Pinker, *How the Mind Works* (New York: W. W. Norton, 1997), 543.

2. Office for National Statistics (ONS), United Kingdom, "Mortality Statistics: Deaths Registered in England and Wales (Series DR)," 2010, www.ons.gov.uk/ons/rel/vsob1/mortality-statistics--deaths-registered-in-england-and-wales--series-dr-/2010/index.html.

CHAPTER 1: THE BEGINNING

1. Steve Chesley, Paul Chodas, and Don Yeomans, "Asteroid 2008 TC3 Strikes Earth: Predictions and Observations Agree," November 4, 2008, National Aeronautics and Space Administration (NASA), http://neo.jpl.nasa.gov/news/2008tc3.html.

2. R. A. Howard, "Microrisks for Medical Decision Analysis," *International Journal of Technology Assessment in Health Care* 5, no. 3 (1989): 357–370.

3. Office for National Statistics (ONS), United Kingdom, "Mortality Statistics: Deaths Registered in England and Wales (Series DR)," www.ons.gov.uk/ons/rel/vsob1/mortality-statistics--deaths-registered-in-england-and-wales--series-dr-/2010/index.html.

4. Sherry L. Murphy, Jiaquan Xu, and Kenneth D. Kochanek, "Deaths: Final Data for 2010," National Vital Statistics Reports, vol. 61, no. 4 (Hyattsville, MD: National Center for Health Statistics, 2013).

5. American Society of Anesthesiologists, "Anesthesia Fast Facts," cited November 4, 2013, www.asahq.org/For-the-Public-and-Media/Press-Room/Anesthesia-Fast-Facts.aspx.

6. US Department of Transportation, "Treatment of the Economic Value of a Statistical Life in Departmental Analyses—2011 Interim Adjustment," 2011, www.dot.gov/policy/transportation-policy/treatment-economic-value-statistical-life, cited November 18, 2013; Department for Transport GMH, United Kingdom, "TAG Unit 3.4: The Safety Objective," Transport

Analysis Guidance—WebTAG, www.dft.gov.uk/webtag/documents/expert /unit3.4.1.php.

CHAPTER 2: INFANCY

1. Centers for Disease Control and Prevention (CDC), "Life Tables," 2013, www.cdc.gov/nchs/products/life_tables.htm, cited November 24, 2013; Office for National Statistics (ONS), United Kingdom, "UK Interim Life Tables, 1980–82 to 2008–10," 2011,

2. J. Schellekens, "Economic Change and Infant Mortality in England, 1580–1837," *Journal of Interdisciplinary History* 32, no. 1 (2001): 1–13.

3. Sherry L. Murphy, Jiaquan Xu, and Kenneth D. Kochanek, "Deaths: Final Data for 2010," National Vital Statistics Reports, vol. 61, no. 4 (Hyattsville, MD: National Center for Health Statistics, 2013); Office for National Statistics (ONS), United Kingdom, "Child Mortality Statistics: Childhood, Infant and Perinatal," 2012, www.ons.gov.uk/ons/rel/vsob1/child-mortality-statistics--childhood--infant-and-perinatal/2010/index.html.

4. National Perinatal Mortality Unit, "The Birthplace Cohort Study: Key Findings," 2012, https://www.npeu.ox.ac.uk/birthplace/results.

5. Office for National Statistics (ONS), United Kingdom, "Unexplained Deaths in Infancy: England and Wales," 2009, www.ons.gov.uk/ons/rel/child -health/unexplained-deaths-in-infancy--england-and-wales/2009/new -component.html.

6. UN Inter-Agency Group for Child Mortality Estimation, "Child Mortality Estimates," 2013, www.childmortality.org/.

7. UNICEF, Statistics by Area / Child Survival and Health, "Trends in Infant Mortality Rates, 1960–2012," 2012, www.childinfo.org/mortality_imrcountry data.php.

8. Centers for Disease Control and Prevention (CDC), "Health, United States," "Table 13: Infant Mortality Rates, Fetal Mortality Rates, and Perinatal Mortality Rates, by Race: United States, Selected Years, 1950–2010," 2012, ftp://ftp.cdc.gov/pub/Health_Statistics/NCHS/Publications/Health_US/hus 12tables/table013.xls, cited November 24, 2013.

9. United Nations Development Program, Millennium Development Goals (MDG), MDG Monitor, Goal 4: "Reduce Child Mortality," www.mdgmonitor .org/goal4.cfm.

10. J. M. Rudski, W. Osei, A. R. Jacobson, and C. R. Lynch, "Would You Rather Be Injured by Lightning or a Downed Power Line? Preference for Natural Hazards," *Judgment and Decision Making* 6, no. 4 (2011): 314–322.

CHAPTER 3: VIOLENCE

1. Federal Bureau of Investigation (FBI), "Crime in the United States, 2012," Expanded Homicide Data, "Table 9: Murder Victims by Age [and] by Weapon," www.fbi.gov/about-us/cjis/ucr/crime-in-the-u.s/2012/crime

-in-the-u.s.-2012/offenses-known-to-law-enforcement/expanded-homicide /expanded_homicide_data_table_9_murder_victims_by_age_by_weapon _2012.xls; Home Office, United Kingdom, "Homicides, Firearm Offences and Intimate Violence, 2010 to 2011: Supplementary Volume 2 to Crime in England and Wales, 2010 to 2011" 2012, https://www.gov.uk/government/publications /homicides-firearm-offences-and-intimate-violence-2010-to-2011-supple mentary-volume-2-to-crime-in-england-and-wales-2010-to-2011.

2. FBI, "Crime in the United States, 2012: Uniform Crime Report," www .fbi.gov/about-us/cjis/ucr/crime-in-the-u.s/2012/crime-in-the-u.s.-2012 /violent-crime/murder/murdermain.pdf.

3. Alexia Cooper and Erica L. Smith, "Homicide Trends in the United States, 1980–2008: Annual Rates for 2009 and 2010," November 2011, US Depart- ment of Justice, www.bjs.gov/content/pub/pdf/htus8008.pdf.

4. National Incidence Studies of Missing, Abducted, Runaway and Throw- away Children (NISMART), 2002, https://www.ncjrs.gov/html/ojjdp/nismart /04/. The 2002 study was known as NISMART 2. NISMART 3 was due to be published in 2013. There have been later estimates as well, such as those by Richard J. Estes at the University of Pennsylvania.

5. J. Robert Flores, "National Estimates of Missing Children: An Overview," October 2002, National Incidence Studies of Missing, Abducted, Runaway, and Throwaway Children, US Department of Justice, Office of Justice Programs, Office of Juvenile Justice and Delinquency Prevention, https://www.ncjrs.gov /pdffiles1/ojjdp/196465.pdf.

6. G. Newiss, "Child Abduction: Understanding Police Recorded Crime Statis- tics," 2008, Child and Maternal Health Intelligence Network, www.chimat.org .uk/resource/item.aspx?RID=62767.

7. National Society for the Prevention of Cruelty to Children (NSPCC), "Statistics on Child Sexual Abuse, October 2013, www.nspcc.org.uk/Inform /resourcesforprofessionals/sexualabuse/statistics_wda87833.html.

8. Stanley Cohen, *Folk Devils and Moral Panics*, 30th anniversary ed. (London: Routledge, 2002).

CHAPTER 4: NOTHING

1. "Pancreatic Cancer Risk Increases with Every 2 Strips of Bacon You Eat," CBS News, January 13, 2013, www.cbsnews.com/8301-504763_162-57358898 -10391704/pancreatic-cancer-risk-increases-with-every-2-strips-of-bacon-you -eat-study/; S. C. Larsson and A. Wolk, "Red and Processed Meat Consumption and Risk of Pancreatic Cancer: Meta-Analysis of Prospective Studies," *British Journal of Cancer* 106, (2012): 603–607.

2. Kate Devlin, "Nine in 10 People Carry Gene Which Increases Chance of High Blood Pressure," *Daily Telegraph*, February 15, 2009, www.telegraph.co.uk /health/healthnews/4630664/Nine-in-10-people-carry-gene-which-increases -chance-of-high-blood-pressure.html.

3. Virginia Woolf, *The Common Reader*, 1st ser., annotated ed. (Orlando, FL: Houghton Mifflin Harcourt, 2002).

4. Daniel Kahneman, *Thinking, Fast and Slow* (New York: Farrar, Straus and Giroux, 2011).

5. Roger Harrabin, Anna Coote, and Jessica Allen, *Health in the News: Risk, Reporting and Media Influence* (London: King's Fund, 2003).

CHAPTER 5: ACCIDENTS

1. Office for National Statistics (ONS), United Kingdom, "Mortality Statistics: Deaths Registered in England and Wales (Series DR)," 2011, www.ons.gov.uk/ons/rel/vsob1/mortality-statistics--deaths-registered-in-england-and-wales--series-dr-/2010/index.html; Sherry L. Murphy, Jiaquan Xu, and Kenneth D. Kochanek, "Deaths: Final Data for 2010," National Vital Statistics Reports, vol. 61, no. 4 (Hyattsville, MD: National Center for Health Statistics, 2013).

2. Nick Britten, "Girl Cannot Walk to Bus Stop Alone," *Daily Telegraph*, September 14, 2010, www.telegraph.co.uk/news/uknews/8001444/Girl-cannot-walk-to-bus-stop-alone.html.

3. US Department of Transportation, National Highway Traffic Safety Administration, "Traffic Safety Facts: 2010 Data—Pedestrians," August 2012, www-nrd.nhtsa.dot.gov/Pubs/811625.PDF, cited November 24, 2013.

4. L. J. Savage, "The Theory of Statistical Decision," *Journal of the American Statistical Association* 46, no. 253 (1951): 55–67.

5. Office for National Statistics (ONS), United Kingdom, "Avoidable Mortality in England and Wales," 2012, www.ons.gov.uk/ons/rel/subnational-health4/avoidable-mortality-in-england-and-wales/2010/index.html; Sherry L. Murphy, Jiaquan Xu, and Kenneth D. Kochanek, "Deaths: Final Data for 2010," National Vital Statistics Reports, vol. 61, no. 4 (Hyattsville, MD: National Center for Health Statistics, 2013).

6. Central Statistical Office, United Kingdom, *Annual Statistical Abstract 1951*; J. Moran, "Crossing the Road in Britain, 1931–1976," *Historical Journal* 49, no. 2 (2006): 477–496.

7. Department for Transport, United Kingdom, "Reported Road Casualties Great Britain: Main Results 2010," June 30, 2011, www.dft.gov.uk/statistics/releases/reported-road-casualties-gb-main-results-2010; Centers for Disease Control and Prevention (CDC), "Health, United States," "Table 33: Death Rates for Motor Vehicle–Related Injuries, by Sex, Race, Hispanic Origin, and Age: United States, Selected Years, 1950–2010," 2012, available at www.cdc.gov/nchs/hus/contents2012.htm#033, cited November 24, 2013.

8. Office for National Statistics (ONS), United Kingdom, "Mortality Statistics: Deaths Registered in England and Wales (Series DR)," 2011, http://www.ons.gov.uk/ons/rel/vsob1/mortality-statistics--deaths-registered-in-england-and-wales--series-dr-/index.html.

9. Murphy et al., "Deaths: Final Data for 2010."

10. Christine Simmons, "Ikea Recalls Over 3 Million Window Blinds, Shades," *The Guardian*, June 10, 2010, www.guardian.co.uk/world/feedarticle /9121407.

11. T. Gill, *No Fear: Growing Up in a Risk Averse Society* (London: Calouste Gulbenkian Foundation, 2007).

12. Health and Safety Executive (HSE), United Kingdom, "Children's Play and Leisure—Promoting a Balanced Response," 2012, www.hse.gov.uk/entertain ment/childrens-play-july-2012.pdf.

13. Countryside Alliance, "Outdoor Education—the Countryside as a Classroom," www.countryside-alliance.org/ca/campaigns-education/give-children -the-opportunity-to-learn-outside-of-the-classroom; Health and Safety Executive (HSE), United Kingdom, "HSE—School Trips—Glenridding Beck— 10 Vital Questions, 2005, www.hse.gov.uk/services/education/school-trips .htm#statistics.

14. Sarah Rainey, "Kellogg's Adds Vitamin D to Cereal to Fight Rickets," *Daily Telegraph*, October 28, 2011, www.telegraph.co.uk/health/health news/8854634/Kelloggs-adds-vitamin-D-to-cereal-to-fight-rickets.html; T. D. Thacher, P. R. Fischer, P. J. Tebben, R. J. Singh, S. S. Cha, J. A. Maxson, et al., "Increasing Incidence of Nutritional Rickets: A Population-Based Study in Olmsted County, Minnesota," *Mayo Clinic Proceedings* 88, no. 2 (2013): 176–183.

CHAPTER 6: VACCINATION

1. Vaccine Liberation Army, "Armed with Knowledge," 2012, http://vaccine liberationarmy.com/.

2. National Health Service (NHS), United Kingdom, "A Guide to Immunisations up to 13 Months of Age," 2007, www.nhs.uk/Planners/vaccinations /Documents/A%20guide%20to%20immunisations%20up%20to%2013%20 months%20of%20age.pdf.

3. Tammy Carrington, "A Vaccination Horror Story," n.d., website of Lawrence Wilson, MD, www.drlwilson.com/articles/VACCINE%20HORROR.htm.

4. Philippa Roxby, "Measles Outbreak Prompts Plea to Vaccinate Children," BBC News, May 27, 2011, www.bbc.co.uk/news/health-13561766.

5. Centers for Disease Control and Prevention (CDC), "School and Childcare Vaccination Surveys," 2012, http://www2a.cdc.gov/nip/schoolsurv/schImm Rqmt.asp.

6. Public Health England, "Measles Notifications and Deaths in England and Wales, 1940–2008," 2012, www.hpa.org.uk/web/HPAweb&HPAwebStandard /HPAweb_C/1195733835814; Centers for Disease Control and Prevention (CDC), "Pinkbook: Measles Chapter—Epidemiology of Vaccine-Preventable Diseases," 2012, www.cdc.gov/vaccines/pubs/pinkbook/meas.html #complications.

7. C. De Martel, J. Ferlay, S. Franceschi, J. Vignat, F. Bray, D. Forman, et al., "Global Burden of Cancers Attributable to Infections in 2008: A Review and

Synthetic Analysis," *The Lancet Oncology* 13, no. 6 (2012): 607–615, www.thelancet .com/journals/lanonc/article/PIIS1470-2045(12)70137-7/abstract.

8. Centers for Disease Control and Prevention (CDC), "Vaccination Coverage Among Children in Kindergarten—United States, 2011–12 School Year," 2012, www.cdc.gov/mmwr/preview/mmwrhtml/mm6133a2.htm, cited November 24, 2013; Centers for Disease Control and Prevention (CDC), "Health, United States," "Table 39: Selected Notifiable Disease Rates and Number of New Cases: United States, Selected Years 1950–2010," 2012, available at www.cdc.gov/nchs /hus/contents2012.htm#039, cited November 24, 2013.

9. Medicines and Healthcare Products Regulatory Agency (MHRA), "Human Papillomavirus (HPV) Vaccine," 2012, www.mhra.gov.uk/PrintPreview /DefaultSplashPP/CON023340?ResultCount=10&DynamicListQuery=& DynamicListSortBy=xCreationDate&DynamicListSortOrder=Desc&Dynamic ListTitle=&PageNumber=1&Title=Human%20papillomavirus%20(HPV)%20 vaccine; Centers for Disease Control and Prevention (CDC), "Vaccination Side Effects: HPV-Cervarix," 2012, www.cdc.gov/vaccines/vac-gen/side-effects.htm #hpvcervarix.

10. Daniel Martin, "NHS Trust Suspends Cervical Cancer Vaccinations After Girl, 14, Dies Within Hours of Jab," *Daily Mail*, October 2, 2009, www.daily mail.co.uk/news/article-1216714/Schoolgirl-14-dies-given-cervical-cancer -jab.html; Neil Durham, "Malignant Tumour Caused HPV Jab Girl's Death," GP Online, October 1, 2009, www.gponline.com/News/article/942531/ Malignant-tumour-caused-HPV-jab-girls-death/.

11. A. S. Goldman, E. J. Schmalstieg, D. H. Freeman, Jr., D. A. Goldman, and F. C. Schmalstieg, Jr., "What Was the Cause of Franklin Delano Roosevelt's Paralytic Illness?" *Journal of Medical Biography* 11, no. 4 (2003): 232–240; Centers for Disease Control and Prevention (CDC), "Seasonal Influenza (Flu): Guillain-Barré Syndrome (GBS)—Questions and Answers," www.cdc.gov/flu/protect /vaccine/guillainbarre.htm.

12. D. J. Sencer and J. D. Millar, "Reflections on the 1976 Swine Flu Vaccination Program," *Emerging Infectious Diseases* 12, no. 1 (2006): 23–28.

13. N. Andrews, J. Stowe, R. Al-Shahi Salman, E. Miller, "Guillain-Barré Syndrome and H1N1 (2009) Pandemic Influenza Vaccination Using an AS03 Adjuvanted Vaccine in the United Kingdom: Self-Controlled Case Series," *Vaccine* 29, no. 45 (2011): 7878–7882.

14. Centers for Disease Control and Prevention (CDC), "Mercury and Thimerosal—Vaccine Safety," 2012, www.cdc.gov/vaccinesafety/Concerns/ thimerosal/.

15. World Health Organization (WHO), "Measles," 2012, www.who.int /mediacentre/factsheets/fs286/en/index.html.

CHAPTER 7: COINCIDENCE

1. "Wasted Stamp," 2012, Understanding Uncertainty, http://understanding uncertainty.org/user-submitted-coincidences/wasted-stamp.

2. David Lodge, *The Art of Fiction* (London: Random House, 2011).

3. "Cambridge Coincidences Collection," 2012, Understanding Uncertainty, http://understandinguncertainty.org/coincidences.

4. P. Diaconis and F. Mosteller, "Methods for Studying Coincidences," *Journal of the American Statistical Association* 84, no. 408 (1989): 853–861; "Biological Daughter," 2012, Understanding Uncertainty, http://understandinguncertainty.org/user-submitted-coincidences/biological-daughter.

5. "Born in the Same Bed," 2012, Understanding Uncertainty, http:/understandinguncertainty.org/user-submitted-coincidences/born-same-bed; "Junk Shop Find," 2012, Understanding Uncertainty, http://understandinguncertainty.org/user-submitted-coincidences/junk-shop-find-0; "Army Coat Hanger," 2012, Understanding Uncertainty, http://understandinguncertainty.org/user-submitted-coincidences/army-coat-hanger.

6. Arthur Koestler, *The Case of the Midwife Toad*, illustrated ed. (London: Hutchinson, 1971).

7. See Diaconis and Mosteller, "Methods for Studying Coincidences."

8. US Department of Commerce, US Census Bureau, "Families and Living Arrangements," "Table F1: Family Households, by Type, Age of Own Children, Age of Family Members, and Age, Race and Hispanic Origin of Householder," 2011, www.census.gov/hhes/families/data/cps2011.html; "Couple Gives Birth to Three Children on the Same Day . . . 14 Years Apart," *Daily Mail*, February 2008, www.dailymail.co.uk/news/article-518525/Couple-gives-birth-children-day--14-years-apart.html; "Mother's Three Children Share Same Birthday," BBC, February 11, 2010 http://news.bbc.co.uk/1/hi/wales/8511586.stm; Andrew Levy, "Happy Birthday to You . . . and You . . . and You Too: Couple's Three Children Born on Same Date," *Daily Mail*, October 13, 2010, www.dailymail.co.uk/news/article-1320113/Happy-birthday-Couple-3-children-born-date.html.

9. "Public Phone Box," 2012, Understanding Uncertainty, http://understandinguncertainty.org/user-submitted-coincidences/public-phone-box.

CHAPTER 8: SEX

1. B. Colombo and G. Masarotto, "Daily Fecundability," *Demographic Research*, September 6, 2000, 3, www.demographic-research.org/Volumes/Vol3/5/default.htm.

2. H. Leridon, "Can Assisted Reproduction Technology Compensate for the Natural Decline in Fertility with Age? A Model Assessment," *Human Reproduction* 19, no. 7 (2004): 1548–1553.

3. National Health Service (NHS), United Kingdom, "Clinical Knowledge Summaries: Effectiveness of Contraceptives," 2012, http://cks.nice.org.uk/#azTab.

4. Office for National Statistics (ONS), United Kingdom, "Conception Statistics, England and Wales," 2011, www.ons.gov.uk/ons/publications/re-reference-tables.html?edition=tcm%3A77-294336.

5. Stephanie J. Ventura, Sally C. Curtin, Joyce C. Abma, and Stanley K. Henshaw, "Estimated Pregnancy Rates and Rates of Pregnancy Outcomes for the United States, 1990–2008," National Vital Statistics Reports, vol. 60, no. 7 (Hyattsville, MD: National Center for Health Statistics, 2012), www.cdc.gov/nchs/data/nvsr/nvsr60/nvsr60_07.pdf.

6. UNICEF, "A League Table of Teenage Births in Rich Nations," 2001, www.unicef-irc.org/publications/328; Organisation for Economic Co-operation and Development (OECD), "OECD Family Database," "Table SF2.4.D: Adolescent Fertility Rates1980 and 2008," www.oecd.org/social/soc/oecdfamilydatabase.htm.

7. B. Varghese, J. E. Maher, T. A. Peterman, B. M. Branson, and R. W. Steketee, "Reducing the Risk of Sexual HIV Transmission: Quantifying the Per-Act Risk for HIV on the Basis of Choice of Partner, Sex Act, and Condom Use," Sexually Transmitted Disease 29, no. 1 (2002): 38–43.

8. R. Platt, P. A. Rice, and W. M. McCormack, "Risk of Acquiring Gonorrhea and Prevalence of Abnormal Adnexal Findings Among Women Recently Exposed to Gonorrhea," Journal of the American Medical Association 250, no. 23 (1983): 3205–3209; K. K. Holmes, D. W. Johnson, and H. J. Trostle, "An Estimate of the Risk of Men Acquiring Gonorrhea by Sexual Contact with Infected Females," American Journal of Epidemiology 91, no. 2 (1970): 170–174.

9. T. S. Nawrot, L. Perez, N. Künzli, E. Munters, and B. Nemery, "Public Health Importance of Triggers of Myocardial Infarction: A Comparative Risk Assessment," The Lancet 377, no. 9767 (2011): 732–740; R. Blanchard and S. J. Hucker, "Age, Transvestism, Bondage, and Concurrent Paraphilic Activities in 117 Fatal Cases of Autoerotic Asphyxia," British Journal of Psychiatry 159, no. 3 (1991): 371–377.

10. HM Treasury, United Kingdom, "Supplementary Green Book Guidance: Optimism Bias," n.d., UK National Archives, www.gov.uk/government/publications/green-book-supplementary-guidance-optimism-bias.

11. T. Sharot, The Optimism Bias: A Tour of the Irrationally Positive Brain (New York: Random House, 2011).

CHAPTER 9: DRUGS

1. Terry M. Parssinen, Secret Passions, Secret Remedies: Narcotic Drugs in British Society, 1820–1930 (Manchester, UK: Manchester University Press, 1983).

2. Arthur Conan Doyle, The Sign of Four (London: Spencer Blackett, 1890).

3. Narcotics Anonymous UK, Stories of Recovery, available at http://ukna.org/content/mp3s.

4. Beef Torrey and Kevin Simonson, eds., Conversations with Hunter S. Thompson (Jackson: University Press of Mississippi, 2008).

5. Cathryn Kemp, Painkiller Addict: From Wreckage to Redemption—My True Story (London: Hachette, 2012).

6. Martin Barrow, "Scandal of 1m Caught in Tranquilliser Addiction Trap," *The Times*, October 1, 2012, www.thetimes.co.uk/tto/health/news/article 3554304.ece.

7. Leaf Fielding, "Why I've Come to Consider Again the Potential Problems of Cannabis," August 28, 2012, *The Guardian*, www.guardian.co.uk/comment isfree/2012/aug/28/why-changed-mind-about-cannabis.

8. Home Office, United Kingdom, "Drug Misuse Declared: Findings from the 2010/11 British Crime Survey England and Wales," July 28, 2011, https://www .gov.uk/government/publications/drug-misuse-declared-findings-from-the -2010-11-british-crime-survey-england-and-wales--12; US Department of Health and Human Services, Substance Abuse and Mental Health Services Administration, "National Survey on Drug Use and Health," "Illicit Drug Use Tables—1.1–1.92," www.samhsa.gov/data/NSDUH/2012SummNatFindDetTables /DetTabs/NSDUH-DetTabsLOTSect1pe2012.htm#TopOfPage.

9. Ibid.

10. Office for National Statistics (ONS), United Kingdom, "Deaths Related to Drug Poisoning in England and Wales," 2011, www.ons.gov.uk/ons/rel/sub national-health3/deaths-related-to-drug-poisoning/2010/index.html.

11. L. A. King, and J. M. Corkery, "An Index of Fatal Toxicity for Drugs of Misuse," *Human Psychopharmacology* 25, no. 2 (2010): 162–166.

12. Ibid.

13. Margaret Warner, Li Hui Chen, Diane M. Makuc, Robert N. Anderson, and Arialdi M. Miniño, "Drug Poisoning Deaths in the United States, 1980–2008," NCHS Data Brief, no. 81, December 2011, Centers for Disease Control and Prevention (CDC), National Center for Health Statistics (NCHS), www.cdc.gov/nchs/data/databriefs/db81.htm; US Department of Health and Human Services, Substance Abuse and Mental Health Services Administration, "National Survey on Drug Use and Health," "Table 1.1A: Types of Illicit Drug Use in Lifetime, Past Year, and Past Month Among Persons Aged 12 or Older: Numbers in Thousands, 2011 and 2012," www.samhsa.gov/data /NSDUH/2012SummNatFindDetTables/DetTabs/NSDUH-DetTabsSect1peTabs 1to46-2012.htm#Tab1.1A.

14. Warner et al., "Drug Poisoning Deaths in the United States,"; Centers for Disease Control and Prevention (CDC), "Unintentional Drug Poisoning in the United States," July 2010, www.cdc.gov/homeand recreationalsafety/pdf/poison-issue-brief.pdf; US Department of Health and Human Services, Substance Abuse and Mental Health Services Administration, "National Survey on Drug Use and Health," www.samhsa.gov/data /NSDUH/2012SummNatFindDetTables/.

15. Advisory Council on the Misuse of Drugs, United Kingdom, "Cannabis Classification and Public Health," May 7, 2008, https://www.gov.uk /government/publications/acmd-cannabis-classification-and-public-health-2008.

16. D. J. Nutt, L. A. King, and L. D. Phillips, "Drug Harms in the UK: A Multicriteria Decision Analysis," *The Lancet* 376, no. 9752 (2010): 1558–1565.

17. D. J. Nutt, "Equasy—An Overlooked Addiction with Implications for the Current Debate on Drug Harms," *Journal of Psychopharmacology* 23, no. 1 (2009): 3–5. Also available at European Coalition for Just and Effective Drug Policies (ENCOD), www.encod.org/info/equasy-a-harmful-addiction.html.

18. Ibid.

CHAPTER 10: BIG RISKS

1. D. M. Kahan, H. Jenkins-Smith, and D. Braman, "Cultural Cognition of Scientific Consensus," February 7, 2010, SSRN eLibrary, http://papers.ssrn.com/sol3/papers.cfm?abstract_id=1549444.

2. D. M. Kahan, P. Slovac, D. Braman, J. Gastil, and G. L. Cohen, "Affect, Values, and Nanotechnology Risk Perceptions: An Experimental Investigation," Yale Law School, Public Law Working Paper No. 155, March 7, 2007, http://papers.ssrn.com/sol3/papers.cfm?abstract_id=968652##.

3. Ibid.

4. Cabinet Office, United Kingdom, National Risk Register, 2012, https://www.gov.uk/government/publications/national-risk-register-of-civil-emergencies.

5. Centers for Disease Control and Prevention (CDC), "Crisis & Emergency Risk Communication," 2012, www.bt.cdc.gov/cerc/.

6. "2011 Germany E. *coli* O104:H4 outbreak," 2012, http://en.wikipedia.org/wiki/2011_Germany_E._coli_O104:H4_outbreak.

CHAPTER 11: GIVING BIRTH

1. World Health Organization (WHO), "Trends in Maternal Mortality: 1990 to 2010," 2012, www.who.int/reproductivehealth/publications/monitoring/9789241503631/en/index.html.

2. Central Intelligence Agency (CIA), The World Factbook, "Country Comparison: Maternal Mortality Rate," https://www.cia.gov/library/publications/the-world-factbook/rankorder/2223rank.html; World Health Organization (WHO), Millennium Development Goals, "MDG 5: Improve Maternal Health," www.who.int/topics/millennium_development_goals/maternal_health/en/.

3. Irvine Loudon, "Maternal Mortality in the Past and Its Relevance to Developing Countries Today," *American Journal of Clinical Nutrition* 72, no. 1 (2000): 241s–246s, http://ajcn.nutrition.org/content/72/1/241s.full.

4. S. Bird, and C. Fairweather, "Recent Military Fatalities in Afghanistan by Cause and Nationality: Period 15, 5 September 2011 to 22 January 2012," February 1, 2012, Medical Research Council, MRC Biostatics Unit, www.mrc-bsu.cam.ac.uk/Publications/PDFs/PERIOD_15_fatalities_in_Afghanistan_by_cause_and_nationality.pdf.

5. Loudon, "Maternal Mortality."

6. UN Development Program, Human Development Indicators, Statistical Tables from the 2013 Human Development Report "Maternal Mortality Ratio,"

https://data.undp.org/dataset/Maternal-mortality-ratio-deaths-of-women-per
-100-0/4gkx-mq89; CIA, World Factbook.

7. WHO, "Trends in Maternal Mortality."

8. Royal College of Anaesthetists, "Death or Brain Damage," 2009, www.rcoa
.ac.uk/document-store/death-or-brain-damage.

CHAPTER 12: GAMBLING

1. King James Bible (Internet version), www.kingjamesbibleonline.org/.

2. S. E. Fienberg, "Randomization and Social Affairs: The 1970 Draft Lottery," *Science* 171, no. 3968 (1971): 255–261.

3. F. N. David, *Games, Gods, and Gambling: A History of Probability and Statistical Ideas* (New York: Dover Publications, 1998); New Living Translation (Internet version), www.newlivingtranslation.com/.

4. "Elizabeth—July 1588, 1–5," Calendar of State Papers Foreign, Elizabeth, Vol. 22, 1936, British History Online, www.british-history.ac.uk/report.aspx ?compid=74849; Ian Hacking, *The Emergence of Probability*, 2nd ed. (New York: Cambridge University Press, 2006).

5. Girolamo Cardano, *Liber de ludo aleae* (Rome: FrancoAngeli, 2006).

6. David, *Games, Gods, and Gambling*.

7. Gerda Reith, *The Age of Chance: Gambling in Western Culture* (London: Routledge, 2002); Mike Atherton, *Gambling* (London: Hachette, 2007).

8. Vic Marks, "Pakistan Embroiled in No-Ball Betting Scandal Against England," *The Guardian*, August 29, 2010, www.guardian.co.uk/sport/2010 /aug/29/pakistan-cricket-betting-allegations.

9. Gamble Aware, "Gambling Facts and Figures," 2012, www.gambleaware.co .uk/recognise-a-Problem/british-gambling-prevalence-survey.

10. Lester E. Dubins and Leonard J. Savage, *How to Gamble If You Must: Inequalities for Stochastic Processes* (New York: McGraw-Hill, 1965).

11. "The WLA Security Control Standard: 2012," World Lottery Association, https://www.world-lotteries.org/cms/index.php?option=com_content &view=article&id=4374%3Athe-wla-security-control-standard-2012-wla-scs 2012&catid=106%3Asecurity-control-standard-scs&Itemid=100177&lang=en.

12. "Gaming Revenue: 10-Year Trend," 2012, American Gaming Association, www.americangaming.org/industry-resources/research/fact-sheets/ gaming-revenue-10-year-trends.

13. Martin Fricker, "Football Fan Wins £585k from 86p Stake," *Daily Mirror*, November 21, 2011, www.mirrorfootball.co.uk/news/Gambler-wins-585k -for-86p-stake-on-19-match-accumulator-thanks-to-Glen-Johnson-87th -minute-Liverpool-winner-v-Chelsea-article833317.html.

14. "DSM-5 Video Series: Gambling Disorder," American Psychiatric Association, www.psychiatry.org/practice/dsm/dsm5/dsm-5-video-series-gambling-disorder.

15. American Psychiatric Association, *Diagnostic and Statistical Manual of Mental Disorders*, 5th ed. (Arlington, VA: American Psychiatric Association, 2013), section 312.31.

16. Oregon Health Authority, "Problem Gambling Treatment Services," 2013, www.oregon.gov/oha/amh/gambling/05-07ad81.pdf; "Oregon Lottery: Readers Guide to the Series on Problem Gambling," *Oregonian*, November 28, 2013, www.oregonlive.com/politics/index.ssf/2013/11/oregon_lottery_readers _guide_t.html; Central and North West London NHS Foundation Trust: Gambling Treatment Centre London, www.cnwl.nhs.uk/gambling.html.

CHAPTER 13: AVERAGE RISKS

1. "Statistics Reveal Britain's 'Mr and Mrs Average,'" BBC, October 13, 2010, www.bbc.co.uk/news/uk-11534042.

2. Office for National Statistics (ONS), "Statistical Bulletin: 2011 Annual Survey of Hours and Earnings (SOC 2000)," 2011, www.ons.gov.uk/ons/rel/ashe /annual-survey-of-hours-and-earnings/ashe-results-2011/ashe-statistical -bulletin-2011.html.

3. For a fun account of a man's search for the "average American" using statistical data for the United States, see Kevin O'Keefe, *The Average American: The Extraordinary Search for the Nation's Most Ordinary Citizen* (New York: PublicAffairs, 2005), with a 2nd ed. ebook version published in 2012.

CHAPTER 14: CHANCE

1. Douglas Adams, *The Hitchhiker's Guide to the Galaxy* (New York: Random House, 1997).

2. RAND Corporation, with a Foreword by Michael D. Rich, *A Million Random Digits with 100,000 Normal Deviates* (Santa Monica: RAND Corporation, 2001 [1955]).

3. Michael Behar, "Burning Question: Why Are Wildfires Defying Long-Standing Computer Models," *The Atlantic*, September 2012, www.theatlantic .com/magazine/archive/2012/09/burning-question/309057/.

CHAPTER 15: TRANSPORTATION

1. John Sergeant, *Give Me Ten Seconds* (New York: Macmillan, 2001).

2. Edgar Sandoval, Barry Paddock, and Kerry Burke, "Victim Describes Shock of Being Randomly Stabbed by Deranged Woman with Steak Knife on Subway," *New York Daily News*, July 9, 2013, www.nydailynews.com/new-york/ uptown/subway-rider-stabbed-woman-cops-article-1.1392851.

3. G. Currie, A. Delbosc, and S. Mahmoud, "Perceptions and Realities of Personal Safety on Public Transport for Young People in Melbourne," conference paper delivered at the Australasian Transport Research Forum held in Canberra, Australia, 2010.

4. Rail Safety and Standards Board, United Kingdom, "Annual Safety Performance Report 2010/11," 2011, www.rssb.co.uk/SPR/REPORTS/Documents /ASPR%202010-11%20Final.pdf.

5. A. Evans, "Fatal Train Accidents on Britain's Main Line Railways: End of 2010 Analysis," 2011, Centre for Transport Studies, Imperial College London,

https://workspace.imperial.ac.uk/cts/Public/Docs/FTAB2012.pdf; US Department of Transportation, Bureau of Transportation Statistics, "Table 1-40: U.S. Passenger-Miles (Millions)," www.rita.dot.gov/bts/sites/rita.dot.gov.bts/files /publications/national_transportation_statistics/html/table_01_40.html, cited November 24, 2013; US Department of Transportation, Bureau of Transportation Statistics, "Transportation Fatalities by Mode," updated October 2013, www.rita.dot.gov/bts/sites/rita.dot.gov.bts/files/publications/national _transportation_statistics/index.html#chapter_2, cited November 24, 2013.

6. "Georgia Varley Train Fall Death: Christopher McGee Jailed," BBC, November 15, 2012, www.bbc.co.uk/news/uk-england-merseyside-20339630.

7. Rail Safety and Standards Board, United Kingdom, "Annual Safety Performance Report 2010/11," 2011, www.rssb.co.uk/SPR/REPORTS/Documents /ASPR%202010-11%20Final.pdf.

8. I. Savage, "Comparing the Fatality Risks in United States Transportation Across Modes and Over Time," *Research in Transportation Economics* 43, no. 1 (2013): 9–22.

9. J. Wolff, "Risk, Fear, Blame, Shame and the Regulation of Public Safety," *Economics and Philosophy* 22, no. 3 (2006): 409–427.

10. G. Gigerenzer, "Out of the Frying Pan into the Fire: Behavioral Reactions to Terrorist Attacks," *Risk Analysis* 26, no. 2 (2006): 347–351.

11. US Department of Transportation, Bureau of Transportation Statistics, "Transportation Fatalities by Mode."

12. V. Hoorens, "Self-Enhancement and Superiority Biases in Social Comparison," *European Review of Social Psychology* 4, no. 1 (1993): 113–139; I. A. McCormick, F. H. Walkey, and D. E. Green, "Comparative Perceptions of Driver Ability—A Confirmation and Expansion," *Accident Analysis Prevention* 18, no. 3 (1986): 205–208.

13. Organisation for Economic Co-operation and Development (OECD), International Transport Forum, International Traffic Safety Data and Analysis Group, International Road Traffic and Accident Database (IRTAD), "IRTAD Road Safety Annual Report," 2012, available at http://internationaltransportforum .org/irtadpublic/index.html.

14. World Health Organization (WHO), "Global Status Report on Road Safety," 2009, available at www.who.int/violence_injury_prevention/road _safety_status/2009/en/.

15. FIA Foundation for the Automobile and Society, "The Missing Link: Road Traffic Injuries and the Millennium Development Goals," 2010, available at www.fiafoundation.org/publications/Pages/PublicationHome.aspx.

16. WHO, "Global Status Report on Road Safety."

17. R. J. Smeed, "Some Statistical Aspects of Road Safety Research," *Journal of the Royal Statistical Society*, Series A (General), 112, no. 1 (1949): 1.

18. M. Oakes and R. Bor, "The Psychology of Fear of Flying (Part I): A Critical Evaluation of Current Perspectives on the Nature, Prevalence and Etiology

of Fear of Flying," *Travel Medicine and Infectious Disease* 8, no. 6 (2010): 327–338.

19. British Airways, "Flying with Confidence—Fear of Flying Course from British Airways," available at http://flyingwithconfidence.com/.

20. Plane Crash Info, http://planecrashinfo.com/index.html.

21. National Transportation Safety Board (NTSB), Aviation Statistics, www.ntsb.gov/data/aviation_stats.html.

22. Department for Transport, United Kingdom, Aviation—Statistics, https://www.gov.uk/government/organisations/department-for-transport/series/aviation-statistics.

23. National Transportation Safety Board (NTSB), "Preliminary Monthly Summary," 2012, www.ntsb.gov/data/monthly/curr_mo.TXT.

CHAPTER 16: EXTREME SPORTS

1. Tom Goodenough, "Last One on the Ground Is a Rotten Egg! Spectacular Photos of Daredevils Diving in Base-Jumping Race," June 25, 2012, www.dailymail.co.uk/news/article-2164332/World-Base-Race-2012-Spectacular-photos-daredevils-diving-base-jumping-race.html.

2. Ronald Clark, *The Victorian Mountaineers* (London: Batsford, 1953).

3. J. S. Windsor, P. G. Firth, M. P. Grocott, G. W. Rodway, and H. E. Montgomery, "Mountain Mortality: A Review of Deaths That Occur During Recreational Activities in the Mountains," *Postgraduate Medical Journal* 85, no. 1004 (2009): 316–321; A. Pollard and C. Clarke, "Deaths During Mountaineering at Extreme Altitude," *The Lancet* 331, no. 8597 (1988): 1277.

4. "Franz Reichelt," Wikipedia, http://en.wikipedia.org/wiki/Franz_Reichelt. The terrifying film of Reichelt's jump is on YouTube at www.youtube.com/watch?v=BepyTSzueno.

5. United States Parachute Association, "Skydiving Safety," n.d., www.uspa.org/AboutSkydiving/SkydivingSafety/tabid/526/Default.aspx.

6. K. Soreide, C. L. Ellingsen, and V. Knutson, "How Dangerous Is BASE Jumping? An Analysis of Adverse Events in 20,850 Jumps from the Kjerag Massif, Norway," *Journal of Trauma: Injury, Infection, and Critical Care* 62, no. 5 (2007): 1113–1117.

7. British Sub-Aqua Club, "UK Diving Fatalities Review," www.bsac.com/page.asp?section=3780&sectionTitle=UK+Diving+Fatalities+Review.

8. D. A. Redelmeier and J. A. Greenwald, "Competing Risks of Mortality with Marathons: Retrospective Analysis," *British Medical Journal* 335, no. 7633 (2007): 1275–1277; C. Kipps, S. Sharma, and D. T. Pedoe, "The Incidence of Exercise-Associated Hyponatraemia in the London Marathon," *British Journal of Sports Medicine* 45, no. 1 (2011): 14–19.

9. US Consumer Product Safety Commission (CPSC), National Electronic Injury Surveillance System (NEISS), https://www.cpsc.gov/en/Research--Statistics/NEISS-Injury-Data/, cited November 24, 2013.

10. Royal Society for the Prevention of Accidents, "Home and Leisure

Accident Statistics: RoSPA : HASS and LASS," www.hassandlass.org.uk.

11. S. Lyng, *Edgework: the Sociology of Risk-taking* (New York: Routledge, 2004).

12. P. Bennett, K. Calman, S. Curtis, and D. Fischbacher-Smith, *Risk Communication and Public Health* (Oxford: Oxford University Press, 2009).

CHAPTER 17: LIFESTYLE

1. Jo Willey, "Less Meat, More Veg is the Secret for Longer Life," *Daily Express,* March 13, 2012, www.express.co.uk/posts/view/307781; A. Pan, Q. Sun, A. M. Bernstein, M. B. Schulze, J. E. Manson, M. J. Stampfer, et al., "Red Meat Consumption and Mortality: Results from 2 Prospective Cohort Studies," *Archives of Internal Medicine* 172, no. 7 (2012): 555–563.

2. Centers for Disease Control and Prevention (CDC), "Life Tables," 2013, www.cdc.gov/nchs/products/life_tables.htm, November 24, 2013.

3. A. Partington, *The Oxford Dictionary of Quotations* (Oxford: Oxford University Press, 1996).

4. M. Shaw, R. Mitchell, and D. Dorling, "Time for a Smoke? One Cigarette Reduces Your Life by 11 Minutes," *British Medical Journal* 320, no. 7226 (2000): 53.

5. R. Doll, "Mortality in Relation to Smoking: 50 Years' Observations on Male British Doctors," *British Medical Journal* 328 (2004): 1519–1520.

6. D. Spiegelhalter, "Using Speed of Ageing and 'Microlives' to Communicate the Effects of Lifetime Habits and Environment," *British Medical Journal* 345 (2012): e8223.

7. Prospective Studies Collaboration, "Body-Mass Index and Cause-Specific Mortality in 900,000 Adults: Collaborative Analyses of 57 Prospective Studies," *The Lancet* 373 (2009): 1083–1096.

8. K.-T. Khaw, N. Wareham, S. Bingham, A. Welch, R. Luben, and N. Day, "Combined Impact of Health Behaviours and Mortality in Men and Women: The EPIC-Norfolk Prospective Population Study," *PLoS Medicine* 5, no. 1 (2008): e12.

9. National Health Service, United Kingdom, Information Centre, "Statistics on Obesity, Physical Activity and Diet: England," 2012, www.hscic.gov.uk/searchcatalogue?productid=4787&topics=2%2fPublic+health%2f Lifestyle%2fPhysical+activity&sort=Relevance&size=10&page=1; Centers for Disease Control and Prevention (CDC), "Adult Participation in Aerobic and Muscle-Strengthening Physical Activities—United States," 2011, www.cdc.gov/mmwr/preview/mmwrhtml/mm6217a2.htm?s_cid=mm6217a2_w#tab1, cited November 24, 2013.

10. J. Woodcock, O. H. Franco, N. Orsini, and I. Roberts, "Non-Vigorous Physical Activity and All-Cause Mortality: Systematic Review and Meta-Analysis of Cohort Studies," *International Journal of Epidemiology* 40, no. 1 (2011): 121–138.

11. L. Byberg, H. Melhus, R. Gedeborg, J. Sundström, A. Ahlbom, B. Zethelius, et al., "Total Mortality After Changes in Leisure Time Physical Activity in 50 Year Old Men: 35 Year Follow-Up of Population Based Cohort," *British Medical Journal* 338 (2009): b688.

12. Ben Goldacre, "Vitamin Pills Can Lead You to Take Health Risks," *The Guardian*, August 26, 2011, www.guardian.co.uk/commentisfree/2011/aug/26/bad -science-vitamin-pills-lead-you-to-take-risks.

13. K. A. Shaw, H. C. Gennat, P. O'Rourke, and C. Del Mar, "Exercise for Overweight or Obesity," Cochrane Library, October 18, 2006, http://onlinelibrary .wiley.com/doi/10.1002/14651858.CD003817.pub3/abstract.

CHAPTER 18: HEALTH AND SAFETY

1. Health and Safety Executive (HSE), United Kingdom, "Myth: You Can't Throw Out Sweets at Pantos," 2009, www.hse.gov.uk/myth/dec09.htm.

2. Nigel Bunyan, "Health and Safety Fears Are 'Taking the Joy Out of Playtime,'" *Daily Telegraph*, July 1, 2011, www.telegraph.co.uk/education/education news/8612145/Health-and-safety-fears-are-taking-the-joy-out-of-playtime.html.

3. John Adams, Risk in a Hypermobile World, www.john-adams.co.uk/.

4. *Hazards Magazine*, January–March 2012, www.hazards.org/haz117/index.htm.

5. Health and Safety Executive (HSE), United Kingdom, "HSE Statistics: Historical Picture," 2000, www.hse.gov.uk/statistics/history/index.htm.

6. Health and Safety Executive (HSE), United Kingdom, "Self-Reported Work-Related Illness and Workplace Injuries," 2008, www.hse.gov.uk/statistics /lfs/index.htm.

7. Health and Safety Executive (HSE), United Kingdom, "Workplace Fatalities and Injuries Statistics in the EU," 2008, www.hse.gov.uk/statistics/european /index.htm.

8. US Department of Labor, Bureau of Labor Statistics, "Census of Fatal Occupational Injuries (CFOI)—Current and Revised Data, 2012, www.bls.gov/iif /oshcfoi1.htm.

9. US Department of Labor, Bureau of Labor Statistics, "Census of Fatal Occupational Injuries (CFOI)—Current and Revised Data," Table A-2, http://stats .bls.gov/iif/oshwc/cfoi/cftb0269.pdf.

10. International Labour Organisation (ILO), "Global Workplace Deaths Vastly Under-Reported, Says ILO," press release, September 18, 2005, www.ilo .org/global/about-the-ilo/press-and-media-centre/news/WCMS_005176/lang --en/index.htm.

11. International Labour Organization (ILO), "XIX World Congress on Safety and Health at Work—ILO Introductory Report: Global Trends and Challenges on Occupational Safety and Health," September 12, 2011, www.ilo.org/safe work/info/publications/WCMS_162662/lang--en/index.htm.

12. Asian Development Bank, "People's Republic of China: Coal Mine Safety Study. Part II: Review and Analysis of International Experience," 2007, www .adb.org/Documents/Reports/Consultant/39657-PRC/39657-02-PRC-TACR.pdf.

13. Department of Energy and Climate Change, United Kingdom, "Coal Mining Technologies and Production Statistics—the Coal Authority," 2012, http://

coal.decc.gov.uk/en/coal/cms/publications/mining/mining.aspx; "Safety in the Pits," *Hazards Magazine*, October–December 2011, www.hazards.org /deadlybusiness/deadlymines.htm.

14. Ibid.

15. J. Tu, "Coal Mining Safety: China's Achilles' Heel," *China Security* 3 (2007): 36–53; Asian Development Bank, "People's Republic of China."; "Safety in the Pits."

16. S. E. Roberts, "Britain's Most Hazardous Occupation: Commercial Fishing," *Accident Analysis & Prevention* 42, no. 1 (2010): 44–49.

17. "The London Beer Flood of 1814," October 24, 2008, h2g2, http://h2g2 .com/dna/h2g2/A42129876.

18. "Boston Molasses Disaster," 2012, Wikipedia, http://en.wikipedia.org /wiki/Boston_Molasses_Disaster.

19. "Bhopal Disaster," 2012, Wikipedia, http://en.wikipedia.org/wiki /Bhopal_disaster.

20. Health and Safety Executive (HSE), United Kingdom, "HSE Statistics: Historical Picture," 2000, www.hse.gov.uk/statistics/history/index.htm.

21. Centers for Disease Control and Prevention (CDC), "Health, United States," "Table 37: Deaths from Selected Occupational Diseases Among Persons Aged 15 and Over: United States, Selected Years, 1980–2010," 2012, www.cdc .gov/nchs/hus/contents2012.htm#chartbookfigures.

22. M. Albin, V. Horstmann, K. Jakobsson, and H. Welinder, "Survival in Cohorts of Asbestos Cement Workers and Controls," *Occupational and Environmental Medicine* 53, no. 2 (1996): 87–93; B. G. Miller, and M. Jacobsen, "Dust Exposure, Pneumoconiosis, and Mortality of Coalminers," *British Journal of Industrial Medicine* 42, no. 11 (1985): 723–733.

23. Health Safety Executive (HSE), United Kingdom, "Reducing Risks, Protecting People: HSE's Decision-Making Process," 2001, www.hse.gov.uk/risk/theory /r2p2.htm.

24. S. M. Bird, and C. B. Fairweather, "Recent Military Fatalities in Afghanistan by Cause and Nationality: Period 15, 5 September 2011 to 22 January 2012," www.mrc-bsu.cam.ac.uk/Publications/PDFs/PERIOD_15_fatalities_in _Afghanistan_by_cause_and_nationality.pdf.

CHAPTER 19: RADIATION

1. John Adams, Risk in a Hypermobile World, www.john-adams.co.uk/.

2. P. Slovic, "Perception of Risk," *Science* 236, no. 4799 (1987): 280–285.

3. P. Mehta and R. Smith-Bindman, "Airport Full-Body Screening: What Is the Risk?" *Archives of Internal Medicine* 171, no. 12 (2011):1112–1115.

4. National Academy of Sciences, "Health Effects of Radiation: Findings of the Radiation Effects Research Foundation," 2003, http://dels.nas.edu/global /nsrb/rerf.

5. United Nations Scientific Committee on the Effects of Atomic Radiation (UNSCEAR), "Assessments of the Chernobyl Accident," 2012, www.unscear.org /unscear/en/chernobyl.html.

6. M.P. Little, D. G. Hoel, J. Molitor, J. D. Boice, R. Wakeford, and C. R. Muirhead, "New Models for Evaluation of Radiation-Induced Lifetime Cancer Risk and Its Uncertainty Employed in the UNSCEAR 2006 Report," *Radiation Research* 169, no. 6 (2008): 660–676.

7. A. Berrington de Gonzalez, M. Mahesh, K.-P. Kim, M. Bhargavan, R. Lewis, F. Mettler, et al., "Projected Cancer Risks from Computed Tomographic Scans Performed in the United States in 2007," *Archives of Internal Medicine* 169, no. 22 (2009): 2071–2077.

8. EU Business News, "After Japan 'Apocalypse,' EU Agrees Nuclear 'Stress Tests,'" March 15, 2011, www.eubusiness.com/news-eu/japan-quake -nuclear.93d.

CHAPTER 20: SPACE

1. Kim Willsher, "Comette Family Home Damaged by Egg-Sized Meteorite," *The Guardian*, October 10, 2011, www.guardian.co.uk/world/2011/oct/10 /comette-family-home-damaged-meteorite.

2. G. Woo, *Calculating Catastrophe* (London: Imperial College Press, 2011); Risk Management Solutions, "Comet and Asteroid Risk: An Analysis of the 1908 Tunguska Event," 2009, www.rms.com/publications/1908_tunguska_event.pdf.

3. National Research Council, *Defending Planet Earth: Near-Earth Object Surveys and Hazard Mitigation Strategies* (Washington, DC: National Academies Press, 2010).

4. M. B. E. Boslough and D. A. Crawford, "Low-Altitude Airbursts and the Impact Threat," *International Journal of Impact Engineering* 35, no. 12 (2008): 1441–1448.

5. S. N. Ward and E Asphaug, "Asteroid Impact Tsunami: A Probabilistic Hazard Assessment," *Icarus* 145, no. 1 (2000): 64–78.

6. Woo, *Calculating Catastrophe*.

7. P. G. Brown, J. D. Assink, L. Astiz, R. Blaauw, M. B. Boslough, J. Borovicka, et al., "A 500-Kiloton Airburst over Chelyabinsk and an Enhanced Hazard from Small Impactors," Letter, *Nature* 503 (2013), www.nature.com/nature /journal/v503/n7475/full/nature12741.html.

8. National Aeronautics and Space Administration (NASA), Near-Earth Object (NEO) Program, http://neo.jpl.nasa.gov/index.html.

9. NASA, NEO Program, "The Torino Impact Hazard Scale," http://neo.jpl.nasa .gov/torino_scale.html.

10. NASA, NEO Program, "Predicting Apophis' Earth Encounters in 2029 and 2036," http://neo.jpl.nasa.gov/apophis/.

11. Discovery News, "Hayabusa Asteroid Probe Awarded World Record," June 20, 2011, http://news.discovery.com/space/hayabusa-asteroid -probe-gets-guinness-world-record-110620.html.

CHAPTER 21: UNEMPLOYMENT

1. Nassim Nicholas Taleb, *The Black Swan: The Impact of the Highly Improbable* (New York: Random House, 2007).

2. US Department of Labor, Bureau of Labor Statistics, "Job Openings and Labor Turnover," 2013, www.bls.gov/news.release/jolts.toc.htm, cited November 24, 2013.

3. Paul Slovic, *The Perception of Risk* (London: Routledge, 2000).

4. Sir Keith Thomas, *The Oxford Book of Work* (Oxford: Oxford University Press, 1999).

5. Trades Union Congress (TUC), "The Costs of Unemployment," 2010, www.tuc.org.uk/sites/default/files/extras/costsofunemployment.pdf.

6. C. J. Ruhm, "Are Recessions Good for Your Health?" *Quarterly Journal of Economics* 115, no. 2 (2000): 617–650; National Institute for Health and Clinical Excellence, United Kingdom, "Worklessness and Health: What Do We Know About the Causal Relationship?" 2005, www.nice.org.uk/niceMedia/documents/work lessness_health.pdf; D. J. Roelfs, E. Shor, K. W. Davidson, and J. E. Schwartz, "Losing Life and Livelihood: A Systematic Review and Meta-Analysis of Unemployment and All-Cause Mortality," *Social Science and Medicine* 72, no. 6 (2011): 840–854.

7. T. Clemens, P. Boyle, and F. Popham, "Unemployment, Mortality and the Problem of Health-Related Selection: Evidence from the Scottish and England & Wales (ONS) Longitudinal Studies," *Health Statistics Quarterly* 43 (2009): 7–13.

8. P. Gregg and E. Tominey, "The Wage Scar from Youth Unemployment," Department of Economics, University of Bristol, Report No. 04/097, 2004, http://ideas.repec.org/p/bri/cmpowp/04-097.html.

9. D. N. F. Bell and D. G. Blanchflower, "Youth Unemployment: Déjà Vu?" Institute for the Study of Labor (IZA), Report No. 4705, 2010, http://ideas.repec.org/p/iza/izadps/dp4705.html.

CHAPTER 22: CRIME

1. US Department of Justice, Bureau of Justice Statistics, "Criminal Victimization, 2012," 2013, www.bjs.gov/index.cfm?ty=pbdetail&iid=4781, cited November 24, 2013.

2. Home Office, United Kingdom, "Crime in England and Wales: Quarterly Update to September 2011," January 2012, https://www.gov.uk/government/publications/crime-in-england-and-wales-quarterly-update-to-september-2011; M. Warr, "Fear of Crime in the United States: Avenues for Research and Policy," 2000, https://www.ncjrs.gov/App/publications/abstract.aspx?ID=185545, cited November 24, 2013.

3. Daniel Kahneman, *Thinking, Fast and Slow* (New York: Farrar, Straus and Giroux, 2011).

4. P. Slovic P., "If I Look at the Mass I Will Never Act: Psychic Numbing and Genocide," in S. Roeser, ed., *Emotions and Risky Technologies* (Dordrect: Springer, 2010), 37–59, http://link.springer.com/chapter/10.1007/978-90-481-8647-1_3.

5. A. Fagerlin, C. Wang, and P. A. Ubel, "Reducing the Influence of Anecdotal Reasoning on People's Health Care Decisions: Is a Picture Worth a Thousand Statistics?" *Medical Decision Making* 25, no. 4 (2005): 398–405.

6. James Wood, *How Fiction Works* (London: Cape, 2008).

7. Home Office, United Kingdom, "Homicides, Firearm Offences and Intimate Violence 2010/11: Supplementary Vol. 2 to Crime in England and Wales," 2012, https://www.gov.uk/government/publications/homicides-firearm -offences-and-intimate-violence-2010-to-2011-supplementary-volume-2-to -crime-in-england-and-wales-2010-to-2011.

8. "Brown Pledges to Tackle Stabbings," BBC News, July 11, 2008, http://news .bbc.co.uk/1/hi/uk/7502569.stm.

9. Jim Gold, "Police Fear War on Cops," NBC News, www.nbcnews.com/id /41235743/#.UoKoRqU1H98.

10. D. Spiegelhalter and A. Barnett, "London Murders: A Predictable Pattern?" *Significance* 6, no. 1 (2009): 5–8.

11. Home Office, "Crime in England and Wales: Quarterly Update to September 2011."

12. Ibid.

13. Federal Bureau of Investigation (FBI), "Table 16: Number of Crimes per 100,000 Inhabitants," 2013, www.fbi.gov/about-us/cjis/ucr/crime-in-the -u.s/2012/crime-in-the-u.s.-2012/tables/16tabledatadecpdf/table_16_rate_by _population_group_2012.xls, cited November 24, 2013.

CHAPTER 23: SURGERY

1. D. Isaacs and D. Fitzgerald, "Seven Alternatives to Evidence Based Medicine," *British Medical Journal* 319, no. 7225 (1999): 1618.

2. G. C. J. Guyatt, "Evidence-Based Medicine: A New Approach to Teaching the Practice of Medicine," *Journal of the American Medical Association* 268, no. 17 (1992): 2420–2425; J. P. A. Ioannidis, "Why Most Published Research Findings Are False," *PLoS Medicine* 2, no. 8 (2005): e124.

3. Atul Gawande, *Complications: A Surgeon's Notes on an Imperfect Science* (New York: Henry Holt, 2002).

4. Charles G. Gross, *A Hole in the Head: More Tales in the History of Neuroscience* (Cambridge, MA: MIT Press, 2009).

5. Davy, S. H. "Researches, Chemical and Philosophical; Chiefly Concerning Nitrous Oxide: Or Dephlogisticated Nitrous Air, and Its Respiration" (St. Paul's Church-yard, UK: J. Johnson, 1800).

6. T. G. Weiser, S. E. Regenbogen, K. D. Thompson, A. B. Haynes, S. R. Lipsitz, W. R. Berry, et al., "An Estimation of the Global Volume of Surgery: A Modelling Strategy Based on Available Data," *The Lancet* 372, no. 9633 (2008): 139–144; American Society of Anesthesiologists, "Anesthesia Fast Facts," 2013, www .asahq.org/For-the-Public-and-Media/Press-Room/Anesthesia-Fast-Facts.aspx.

7. Centers for Disease Control and Prevention (CDC), National Center for Health Statistics (NCHS), "Trends in Inpatient Hospital Deaths: National

Hospital Discharge Survey, 2000–2010," 2013, www.cdc.gov/nchs/data/data briefs/db118.htm, cited November 4, 2013.

8. Florence Nightingale, *Notes on Hospitals* (London: Longman, Green, Longman, Roberts and Green, 1863).

9. E. A. Codman, *A Study in Hospital Efficiency: As Demonstrated by the Case Report of the First Five Years of a Private Hospital* (Boston: T. Todd, 1918).

10. T. B. Ferguson, Jr., B. G. Hammill, E. D. Peterson, E. R. DeLong, and F. L. Grover, "A Decade of Change—Risk Profiles and Outcomes for Isolated Coronary Artery Bypass Grafting Procedures, 1990–1999: A Report from the STS National Database Committee and the Duke Clinical Research Institute," Society of Thoracic Surgeons, *Annals of Thoracic Surgery* 73, no. 2 (2002): 480–489 (discussion 489–490); Care Quality Commission, "Survival Rates—Heart Surgery in United Kingdom 2008–9," http://bluebook.scts.org.

11. New York State Department of Health, "Cardiovascular Disease Data and Statistics," 2012, www.health.ny.gov/statistics/diseases/cardiovascular/.

12. M. J. Campbell, R. M. Jacques, J. Fotheringham, R. Maheswaran, and J. Nicholl, "Developing a Summary Hospital Mortality Index: Retrospective Analysis in English Hospitals over Five Years," *British Medical Journal* 344 (2012): e1001.

13. N. Hawkes, "Patient Coding and the Ratings Game," *British Medical Journal* 25, no. 340 (2010): c2153.

CHAPTER 24: SCREENING

1. Cancer Research UK, "Breast Screening: Accuracy of Mammography," 2012, www.cancerresearchuk.org/cancer-info/cancerstats/types/breast/screening /Other-Issues/#Accuracy.

2. Margaret McCartney, *The Patient Paradox: Why Sexed Up Medicine Is Bad for Your Health* (London: Pinter and Martin, 2012).

3. P. Mehta, and R. Smith-Bindman, "Airport Full-Body Screening: What Is the Risk?" *Archives of Internal Medicine* 171, no. 12 (2011): 1112–1115.

4. NHS Breast Screening Programme, United Kingdom, "Screening for Breast Cancer in England: Past and Future," 2012, www.cancerscreening.nhs .uk/breastscreen/publications/nhsbsp61.html.

5. M. J. Yaffe, and J. G. Mainprize, "Risk of Radiation-Induced Breast Cancer from Mammographic Screening," *Radiology* 258, no. 1 (2011): 98–105.

6. H. Gilbert Welch, Lisa M. Schwartz, and Steve Woloshin, *Overdiagnosed: Making People Sick in the Pursuit of Health* (Boston: Beacon Press, 2011).

7. Cancer Research UK, "Breast Screening Review," 2012, www.cancer researchuk.org/cancer-info/publicpolicy/ourpolicypositions/symptom _Awareness/cancer_screening/breast-screening-review/breast-screening -review; National Health Service (NHS), United Kingdom, Breast Screening Programme, "NHS Breast Screening: Helping You Decide," www.cancer screening.nhs.uk/breastscreen/publications/ia-02.html.

8. R. J. Ablin, "The Great Prostate Mistake," *New York Times*, March 10, 2010, www. nytimes.com/2010/03/10/opinion/10Ablin.html; "Andrew Lloyd-Webber

Calls for Prostate Cancer Screening," BBC, July 19, 2010, http://news
.bbc.co.uk/democracylive/hi/house_of_lords/newsid_8822000/8822506.stm.

9. Liz Thomas and Sophie Borland, "Andrew Lloyd Webber Reveals Prostate
Cancer Battle Has Left Him Impotent," *Daily Mail*, March 30, 2011, www.daily
mail.co.uk/tvshowbiz/article-1371379/Andrew-Lloyd-Webber-reveals-prostate
-cancer-battle-left-impotent.html.

10. M. W. Vernooij, M. A. Ikram, H. L. Tanghe, A. J. P. E. Vincent, A.
Hofman, G. P. Krestin, et al., "Incidental Findings on Brain MRI in the General
Population," *New England Journal of Medicine* 357, no. 18 (2007): 1821–1828.

11. Cancer Research UK, "Prostate Cancer—UK Incidence Statistics," 2011,
http://info.cancerresearchuk.org/cancerstats/types/prostate/incidence/.

12. Welch et al., *Overdiagnosed*.

13. G. L. Andriole, E. D. Crawford, R. L. Grubb, S. S. Buys, D. Chia, T. R.
Church, et al., "Prostate Cancer Screening in the Randomized Prostate, Lung,
Colorectal, and Ovarian Cancer Screening Trial: Mortality Results after 13
Years of Follow-Up," *Journal of the National Cancer Institute* 104, no. 2 (2012): 125–
132; F. H. Schröder, J. Hugosson, M. J. Roobol, T. L. J. Tammela, S. Ciatto, V.
Nelen, et al., "Prostate-Cancer Mortality at 11 Years of Follow-Up," *New England
Journal of Medicine* 366, no. 11 (2012): 981–990.

14. H. G. Welch and B. A. Frankel, "Likelihood That a Woman with
Screen-Detected Breast Cancer Has Had Her 'Life Saved' by That Screening,"
Archives of Internal Medicine 171, no. 22 (2011): 2043–2046.

15. 23andMe, "Genetic Testing for Health, Disease & Ancestry; DNA Test,"
https://www.23andme.com/.

CHAPTER 25: MONEY

1. Simon Fowler, *Workhouse: The People, the Places, the Life Behind Doors* (Kew, UK:
National Archives, 2007).

2. Charles Booth, "The Aged Poor in England and Wales," 1894, Internet Archive, http://archive.org/stream/agedpoorinengla00bootgoog/agedpoorinengla
00bootgoog_djvu.txt; M. A. Crowther, *The Workhouse System, 1834–1929: The History of an English Social Institution* (London: Routledge, 1983); George Orwell, *Down
and Out in Paris and London: A Novel* (New York: Harcourt, Brace & World, 1961).

3. US Department of Commerce, US Census Bureau, "Table 21: Poverty
Status of the Population by Sex and Age: 2010," https://www.census.gov
/population/age/data/2011comp.html; Sudipto Banerjee, "Time Trends in
Poverty for Older Americans Between 2001–2009," April 1, 2012, *EBRI Notes* 33,
no. 4 (2012), available at Social Science Research Network, http://papers.ssrn
.com/abstract=2046720.

4. For a comprehensive discussion of the financial status of the elderly in the
United Kingdom, see Institute for Fiscal Studies, "The 2008 English Longitudinal
Study of Ageing: Financial Circumstances, Health and Well-Being of the Older
Population in England," 2010, www.ifs.org.uk/elsa/report10/elsa_w4-1.pdf.

5. US Census Bureau, Table 21.

6. Commission on Funding of Care and Support, UK Department of Health, "Fairer Funding for All: The Commission's Recommendations to Government," July 2011, www.dilnotcommission.dh.gov.uk/our-report/.

CHAPTER 26: THE END

1. Human Mortality Database, www.mortality.org/.

2. For US mortality data in this chapter, see US Department of Commerce, US Census Bureau, "Table 104: Expectation of Life at Birth, 1960 to 2008, and Projections, 2010 to 2020," www.census.gov/compendia/statab/cats/births _deaths_marriages_divorces/life_expectancy.html; Centers for Disease Control and Prevention (CDC), "Table A: Deaths, Age-Adjusted Death Rates, and Life Expectancy at Birth, by Race and Sex; and Infant Deaths and Mortality Rates, by Race: United States, Final 2010 and Preliminary 2011," www.cdc .gov/nchs/data/nvsr/nvsr61/nvsr61_06.pdf; Donna L. Hoyert and Jiaquan Xu, "Deaths: Preliminary Data for 2011," National Vital Statistics Reports, vol. 61, no. 6 (Hyattsville, MD: National Center for Health Statistics, 2012), www.cdc .gov/nchs/data/nvsr/nvsr61/nvsr61_06.pdf; Kristen Lewis and Sarah Burd-Sharps, American Human Development Report: The Measure of America, 2013– 2014, Measure of America, Social Science Research Council, www.measure ofamerica.org/wp-content/uploads/2013/07/MOA-III-June-18-FINAL.pdf.

3. Office for National Statistics (ONS), United Kingdom, "Life Expectancy at Birth and at Age 65 by Local Areas in the United Kingdom," 2011, www.ons .gov.uk/ons/rel/subnational-health4/life-expec-at-birth-age-65/2004-06 -to-2008-10/index.html.

4. "Survival Worldwide," 2012, Understanding Uncertainty, http:// understandinguncertainty.org/node/272; World Health Organization (WHO), "World Health Statistics 2011, www.who.int/whosis/whostat/2011/en/; The Human Life-Table Database, www.lifetable.de/.

5. Office for National Statistics (ONS), United Kingdom, "Period and Cohort Life Expectancy Tables," 2011, www.ons.gov.uk/ons/rel/lifetables/period -and-cohort-life-expectancy-tables/2010-based/index.html; Office for National Statistics (ONS), United Kingdom, "What Are the Chances of Surviving to Age 100?" 2012, www.ons.gov.uk/ons/rel/lifetables/historic-and -projected-mortality-data-from-the-uk-life-tables/2010-based/rpt-surviving -to-100.html.

6. United Nations, "Global Issues at the United Nations," www.un.org/en /globalissues/ageing/.

7. Office for National Statistics (ONS), United Kingdom, "Health Expectancies at Birth and at Age 65 in the United Kingdom, 2008–2010," August 29, 2012, www.ons.gov.uk/ons/rel/disability-and-health-measurement /health-expectancies-at-birth-and-age-65-in-the-united-kingdom/2008-10 /stb-he-2008-2010.html?format=print.

8. US Department of Commerce, US Census Bureau, *Statistical Abstract of the United States: 2012*, "Table 72: Persons Living Alone by Sex and Age: 1990 to 2010," www.census.gov/compendia/statab/2012/tables/12s0072.pdf; Centers for Disease Control and Prevention (CDC), "Dementia/Alzheimer's Disease," www.cdc.gov/mentalhealth/basics/mental-illness/dementia.htm.

9. Office for National Statistics (ONS), United Kingdom, "Pension Trends," Chapter 2, "Population Change," 2012, www.ons.gov.uk/ons/rel/pensions /pension-trends/chapter-2--population-change--2012-edition-/index.html; M. Knapp and M. Prince, "Dementia UK," 2007, Alzheimer's Society, http://alzheimers.org.uk/site/scripts/download_info.php?fileID=2.

10. Office for National Statistics (ONS), United Kingdom, "Pension Trends," Chapter 3, "Life Expectancy and Healthy Ageing," 2012, www.ons.gov.uk /ons/rel/pensions/pension-trends/chapter-3--life-expectancy-and-healthy -ageing--2012-edition-/index.html.

11. British Parachute Association, "How Safe?" 2012, www.bpa.org.uk/staysafe /how-safe/.

CHAPTER 27: JUDGMENT DAY

1. Bruno De Finetti, *Theory of Probability* (London: John Wiley, 1974).

ILLUSTRATION SOURCES

Figure 1. Author's calculations based on sources cited in text.

Figure 2. Sherry L. Murphy, Jiaquan Xu, and Kenneth D. Kochanek, "Deaths: Final Data for 2010," National Vital Statistics Reports, vol. 61, no. 4 (Hyattsville, MD: National Center for Health Statistics, 2013), Table E.

Figure 3. Office for National Statistics (ONS), United Kingdom, "Child Mortality Statistics: Childhood, Infant and Perinatal, 2011," "Table 17: Stillbirth and Infant Death Rates: Age at Death, 1921–2011," 2011, available at www.ons.gov.uk/ons/publications/re-reference-tables.html?edition=tcm %3A77-296223; UN Inter-Agency Group for Child Mortality Estimation, "Report 2013: Levels and Trends in Child Mortality," 2013, www.child mortality.org/files_v16/download/UNICEF%202013%20IGME%20child%20 mortality%20Report_Final.pdf.

Figure 4. Federal Bureau of Investigation (FBI), "Crime in the United States, 2012," www.fbi.gov/about-us/cjis/ucr/crime-in-the-u.s/2012/crime-in-the -u.s.-2012/.

Figure 5. S. C. Larsson and A. Wolk, "Red and Processed Meat Consumption and Risk of Pancreatic Cancer: Meta-Analysis of Prospective Studies," British Journal of Cancer 106, (2012): 603–607.

Figure 6. Roger Harrabin, Anna Coote, and Jessica Allen, Health in the News: Risk, Reporting and Media Influence (London: King's Fund, 2003).

Figure 7. Centers for Disease Control and Prevention (CDC), "Health, United States," "Table 33: Death Rates for Motor Vehicle–Related Injuries, by Sex, Race, Hispanic Origin, and Age: United States, Selected Years, 1950–2010," 2012, available at www.cdc.gov/nchs/hus/contents2012.htm#033, cited November 24, 2013.

Figure 8. Sherry L. Murphy, Jiaquan Xu, and Kenneth D. Kochanek, "Deaths: Final Data for 2010," National Vital Statistics Reports, vol. 61, no. 4 (Hyattsville, MD: National Center for Health Statistics, 2013).

Figure 9. Sherry L. Murphy, Jiaquan Xu, and Kenneth D. Kochanek, "Deaths:

Final Data for 2010," National Vital Statistics Reports, vol. 61, no. 4 (Hyatts-ville, MD: National Center for Health Statistics, 2013).

Figure 10. Public Health England, "Measles Notifications and Deaths in En-gland and Wales, 1940–2008," 2012, www.hpa.org.uk/web/HPAweb&HPA webStandard/HPAweb_C/1195733835814; Centers for Disease Control and Prevention (CDC), "Vaccination Coverage Among Children in Kindergarten—United States, 2011–12 School Year," 2012, www.cdc.gov/mmwr/preview /mmwrhtml/mm6133a2.htm, cited November 24, 2013; Centers for Disease Control and Prevention (CDC), "Health, United States," "Table 39: Selected Notifiable Disease Rates and Number of New Cases: United States, Selected Years 1950–2010," 2012, available at www.cdc.gov/nchs/hus/contents2012 .htm#039, cited November 24, 2013.

Figure 11. National Health Service, United Kingdom, "Clinical Knowledge Summaries: Effectiveness of Contraceptives," 2012, www.cks.nhs.uk/clinical _topics/by_alphabet/c.

Figure 12. Centers for Disease Control and Prevention (CDC), "Sexually Trans-mitted Disease Surveillance 2011," "Table 1: Cases of Sexually Transmitted Diseases Reported by State Health Departments and Rates per 100,000 Pop-ulation, United States, 1941–2011," www.cdc.gov/std/stats11/surv2011.pdf.

Figure 13. Health Protection Agency (now part of Public Health England), United Kingdom, 2012, in response to author's request.

Figure 14. US Department of Health and Human Services, Substance Abuse and Mental Health Services Administration, "National Survey on Drug Use and Health," "Table 1.1B: Types of Illicit Drug Use in Lifetime, Past Year, and Past Month Among Persons Aged 12 or Older: Percentages, 2011 and 2012," www.samhsa.gov/data/NSDUH/2012SummNatFindDetTables/DetTabs /NSDUH-DetTabsSect1peTabs1to46-2012.htm.

Figure 15. Cabinet Office, United Kingdom, National Risk Register, 2012, https://www.gov.uk/government/publications/national-risk-register -of-civil-emergencies.

Figure 16. "Historical Mortality Rates of Puerperal Fever," 2012, Wikipedia, http://en.wikipedia.org/w/index.php?title=Historical_mortality_rates_of_ puerperal_fever&oldid=516214953.

Figure 17. UN Development Program, Human Development Indicators, Sta-tistical Tables from the 2013 Human Development Report "Maternal Mortality Ratio," https://data.undp.org/dataset/Maternal-mortality-ratio -deaths-of-women-per-100-0/4gkx-mq89.

Figure 18. Author's graphic based on lottery results. http://lottery.mersey world.com/Winning_index.html.

Figure 19. Plane Crash Info, http://planecrashinfo.com/index.html.

Figure 20. Plane Crash Info, http://planecrashinfo.com/index.html.

Figure 21. Author's calculations based on sources cited in text.

Figure 22. Photo: David Spiegelhalter.

Figure 23. Public Health England, "Dose Comparisons for Ionising Radiation," n.d., www.hpa.org.uk/Topics/Radiation/UnderstandingRadiation/Under standingRadiationTopics/DoseComparisonsForIonisingRadiation/; World Health Organization (WHO), "Preliminary Dose Estimation from the Nuclear Accident After the 2011 Great East Japan Earthquake and Tsunami," 2012, www.who.int/ionizing_radiation/pub_meet/fukushima_dose_assessment /en/index.html.

Figure 24. US Department of Labor, Bureau of Labor Statistics, "Job Openings and Labor Turnover Survey," www.bls.gov/jlt/; for UK labor market flow figures, see Office for National Statistics (ONS), United Kingdom, "Labour Market Flows, November 2011 (Experimental Statistics)," 2011, www.ons .gov.uk/ons/rel/lms/labour-market-statistics/november-2011/art-labour -market-flows--july-to-september-2011.html.

Figure 25. US Department of Labor, Bureau of Labor Statistics, "Job Openings and Labor Turnover Survey," www.bls.gov/jlt/.

Figure 26. Data supplied by Home Office, UK.

Figure 27. "Harold Shipman's Clinical Practice, 1974–1998: A Clinical Audit Commissioned by the Chief Medical Officer," Department of Health, United Kingdom, January 2001.

Figure 28 (A and B). Lynn Langton, Michael Planty, and Jennifer Truman, "Criminal Victimization, 2012," US Department of Justice, Office of Justice Programs, Bureau of Justice Statistics, October 2013, www.bjs.gov/index .cfm?ty=pbdetail&iid=4781.

Figure 29. Lynn Langton, Michael Planty, and Jennifer Truman, "Criminal Victimization, 2012," US Department of Justice, Office of Justice Programs, Bureau of Justice Statistics, October 2013, www.bjs.gov/index .cfm?ty=pbdetail&iid=4781.

Figure 30. New York State Department of Health, "Cardiovascular Disease Data and Statistics," 2012, www.health.ny.gov/statistics/diseases/cardiovascular/.

Figure 31. Cancer Research UK, "Breast Screening Review," 2012, www .cancerresearchuk.org/cancer-info/publicpolicy/ourpolicypositions/symptom _Awareness/cancer_screening/breast-screening-review/breast-screening -review; National Health Service (NHS), United Kingdom, Breast Screening Programme, "NHS Breast Screening: Helping You Decide," www.cancer screening.nhs.uk/breastscreen/publications/ia-02.html.

Figure 32. Supplied privately to author from 23andMe, "Genetic Testing for Health, Disease & Ancestry; DNA Test," https://www.23andme.com/.

Figure 33. US Department of Commerce, US Census Bureau, "Current Population Survey: Annual Social and Economic Supplement, 2011," "Table 21: Poverty Status of the Population by Sex and Age: 2010," www.census.gov/cps/.

Figure 34. Department for Work and Pensions, United Kingdom, "Households Below Average Income," June 2012, http://research.dwp.gov.uk/asd/index .php?page=hbai.

Figure 35. Edmond Halley, "An Estimate of the Degrees of the Mortality of Mankind, Drawn from Curious Tables of the Births and Funerals at the City of Breslaw; With an Attempt to Ascertain the Price of Annuities upon Lives," Royal Society of London, 1753, Internet Archive, http://archive.org/details /philtrans05474358.

Figure 36. Edmond Halley, "An Estimate of the Degrees of the Mortality of Mankind, Drawn from Curious Tables of the Births and Funerals at the City of Breslaw; With an Attempt to Ascertain the Price of Annuities upon Lives," Royal Society of London, 1753, Internet Archive, http://archive.org/details /philtrans05474358.

Figure 37. Sherry L. Murphy, Jiaquan Xu, and Kenneth D. Kochanek, "Deaths: Final Data for 2010," National Vital Statistics Reports, vol. 61, no. 4 (Hyattsville, MD: National Center for Health Statistics, 2013); UN Inter-Agency Group for Child Mortality Estimation, "Child Mortality Estimates," 2012, www .childmortality.org/; US Department of Transportation, National Highway Traffic Safety Administration, "Traffic Safety Facts: 2010 Data—Pedestrians," August 2012, www-nrd.nhtsa.dot.gov/Pubs/811625.PDF; Federal Bureau of Investigation (FBI), "Table 16: Number of Crimes per 100,000 Inhabitants," 2013, www.fbi.gov/about-us/cjis/ucr/crime-in-the-u.s/2012/crime-in-the -u.s.-2012/tables/16tabledatadecpdf/table_16_rate_by_population _group_2012.xls; American Society of Anesthesiologists, "Anesthesia Fast Facts," 2013, www.asahq.org/For-the-Public-and-Media/Press-Room /Anesthesia-Fast-Facts.aspx; New York State Department of Health. "Cardiovascular Disease Data and Statistics," 2012, www.health.ny.gov/statistics /diseases/cardiovascular/; US Department of Labor, Bureau of Labor Statistics, "Census of Fatal Occupational Injuries (CFOI)—Current and Revised Data, 2012, www.bls.gov/iif/oshcfoi1.htm; Congressional Research Service Reports-General National Security. "Troop Levels in the Afghan and Iraq Wars, FY2001-FY2012: Cost and Other Potential Issues," 2012, www .fas.org/sgp/crs/natsec/; "RAF Bomber Command," Wikipedia, http://en.wiki pedia.org/w/index.php?title=RAF_Bomber_Command&oldid=531643454 (accessed January 29, 2013); Department for Transport, United Kingdom, "Reported Road Casualties Great Britain: Main Results, 2010," 2011, www.dft .gov.uk/statistics/releases/reported-road-casualties-gb-main-results -2010; US Department of Transportation, Bureau of Transportation Statistics, "Transportation Fatalities by Mode," updated October 2013, www.rita.dot.gov/bts/sites/rita.dot.gov.bts/files/publications/national _transportation_statistics/index.html#chapter_2; National Transportation Safety Board (NTSB), "Aviation Statistical Reports," www.ntsb.gov/data /aviation_stats.html; British Sub-Aqua Club, "UK Diving Fatalities Review," www.bsac.com/page.asp?section=3780&sectionTitle=UK+Diving+Fatalities +Review; Health Safety Executive (HSE), United Kingdom, "Risk Education—Statistics," 2003, www.hse.gov.uk/education/statistics.htm; United

States Parachute Association, "Skydiving Safety," www.uspa.org/AboutSky diving/SkydivingSafety/tabid/526/Default.aspx; D. A. Redelmeier and J. A. Greenwald, "Competing Risks of Mortality with Marathons: Retrospective Analysis," British Medical Journal 335, no. 7633 (2007):1275–1277; L. A. King and J. M. Corkery, "An Index of Fatal Toxicity for Drugs of Misuse," Human Psychopharmacology 25, no. 2 (2010): 162–166; Home Office, United Kingdom, "Drug Misuse Declared: Findings from the 2010/11 British Crime Survey England and Wales," 2011, www.homeoffice.gov.uk/publications /science-research-statistics/research-statistics/crime-research/hosb1211/; Margaret Warner, Li Hui Chen, Diane M. Makuc, Robert N. Anderson, and Arialdi M. Miniño, "Drug Poisoning Deaths in the United States, 1980– 2008," NCHS Data Brief, no. 81, December 2011, Centers for Disease Control and Prevention (CDC), National Center for Health Statistics (NCHS), www .cdc.gov/nchs/data/databriefs/db81.htm; US Department of Health and Human Services, Substance Abuse and Mental Health Services Administration, "National Survey on Drug Use and Health," "Table 1.1A: Types of Illicit Drug Use in Lifetime, Past Year, and Past Month Among Persons Aged 12 or Older: Numbers in Thousands, 2011 and 2012," www .samhsa.gov/data/NSDUH/2012SummNatFindDetTables/DetTabs/NSDUH -DetTabsSect1peTabs1to46-2012.htm#Tab1.1A. National Research Council, *Defending Planet Earth: Near-Earth Object Surveys and Hazard Mitigation Strategies* (Washington, DC: National Academies Press, 2010); Asian Development Bank. "Peoples Republic of China: Coal Mine Safety Study. Part II: Review and Analysis of International Experience," 2007, www2.adb.org/Documents/ Reports/Consultant/39657-PRC/39657-02-PRC-TACR.pdf; US Department of Labor, Bureau of Labor Statistics, "Census of Fatal Occupational Injuries (CFOI)—Current and Revised Data, 2012, www.bls.gov/iif/oshcfoi1.htm.

Figure 38. R. Doll, "Mortality in Relation to Smoking: 50 Years' Observations on Male British Doctors," British Medical Journal 328 (2004): 1519–1520; A. Di Castelnuovo, S. Costanzo, V. Bagnardi, M. B. Donati, L. Iacoviello, and G. De Gaetano, "Alcohol Dosing and Total Mortality in Men and Women: An Updated Meta-Analysis of 34 Prospective Studies," Archives of Internal Medicine 166, no. 22 (2006): 2437–2345; Prospective Studies Collaboration, "Body-Mass Index and Cause-Specific Mortality in 900,000 Adults: Collaborative Analyses of 57 Prospective Studies," The Lancet 373 (2009): 1083–1096; K. Wijndaele, S. Brage, H. Besson, K. T. Khaw, S. J. Sharp, R. Luben, et al., "Television Viewing Time Independently Predicts All-Cause and Cardiovascular Mortality: The EPIC Norfolk Study," International Journal of Epidemiology 40, no. 1 (2011): 150–159; A. Pan, Q. Sun, A. M. Bernstein, M. B. Schulze, J. E. Manson, M. J. Stampfer, et al. "Red Meat Consumption and Mortality: Results from 2 Prospective Cohort Studies," Archives of Internal Medicine 172, no. 7 (2012): 555–563; K. T. Khaw, N. Wareham, S. Bingham, A. Welch, R. Luben, and N. Day, "Combined Impact of Health Behaviours and Mortality in Men and Women: The EPIC-Norfolk Prospective Population Study," *PLoS Medicine* 5, no. 1 (2008): e12; N. D. Freedman, Y. Park, C. C. Abnet, A. R. Hollenbeck, and R. Sinha, "Association of Coffee Drinking with Total and Cause-Specific Mortality," *New England Journal of Medicine* 366, no. 20 (2012): 1891–1904; J. Woodcock, O. H. Franco, N. Orsini, and I. Roberts, "Non-Vigorous Physical Activity and All-Cause Mortality: Systematic Review and Meta-Analysis of Cohort Studies," *International Journal of Epidemiology* 40, no. 1 (2011): 121–138; K. K. Ray, S. R. K. Seshasai, S. Erqou, P. Sever, J. W. Jukema, I. Ford, et al., "Statins and All-Cause Mortality in High-Risk Primary Prevention: A Meta-Analysis of 11 Randomized Controlled Trials Involving 65,229 Participants," *Archives of Internal Medicine* 170, no. 12 (2010): 1024–1031; C. A. Pope, M. Ezzati, and D. W. Dockery, "Fine-Particulate Air Pollution and Life Expectancy in the United States," *New England Journal of Medicine* 360, no. 4 (2009): 376–386; Office for National Statistics (ONS), United Kingdom, "Interim Life Tables, 2008–10," www.ons.gov.uk/ons/rel/lifetables/interim-life-tables/2008-2010/index.html; Human Mortality Database, 2012, www.mortality.org/; Public Health England, "Dose Comparisons for Ionising Radiation," n.d., www.hpa.org.uk/Topics/Radiation/UnderstandingRadiation/UnderstandingRadiationTopics/DoseComparisonsForIonisingRadiation/.

INDEX